INTEGRATED TRANSPORT POLICY

Integrated Transport Policy

Implications for regulation and competition

Edited by
JOHN PRESTON
Transport Studies Unit, University of Oxford, UK
HELEN LAWTON SMITH
Coventry Business School, Coventry University, UK
DAVID STARKIE
Economics-Plus Limited, UK

Routledge
Taylor & Francis Group

LONDON AND NEW YORK

First published 2000 by Ashgate Publishing

Reissued 2018 by Routledge
2 Park Square, Milton Park, Abingdon, Oxon OX14 4RN
711 Third Avenue, New York, NY 10017, USA

Routledge is an imprint of the Taylor & Francis Group, an informa business

Publisher's Note
The publisher has gone to great lengths to ensure the quality of this reprint but points out that some imperfections in the original copies may be apparent.

Disclaimer
The publisher has made every effort to trace copyright holders and welcomes correspondence from those they have been unable to contact.

A Library of Congress record exists under LC control number: 00134000

ISBN 13: 978-1-138-73729-7 (hbk)
ISBN 13: 978-1-138-73722-8 (pbk)
ISBN 13: 978-1-315-18553-8 (ebk)

Contents

PART IV: INSTITUTIONS: IS THE CURRENT REGULATORY FRAMEWORK ADEQUATE?

List of Contributors

(and position at the time of the seminar)

Chris Bolt, Director, Economic Regulation Group, Office of the Rail Regulator.

Professor Bill Bradshaw, Centre for Socio-Legal Studies, Wolfson College, University of Oxford.

Professor Kenneth Button, Institute of Public Policy, George Mason University.

Dr John Dodgson, Associate Director, NERA.

Melanie Farquharson, Partner, Simmons and Simmons.

Mike Hoban, Competition Policy Division, Office of Fair Trading.

Dr Helen Lawton Smith, Director, Science Policy Studies, Regulatory Policy Research Centre, Hertford College, Oxford.

Professor Chris Nash, Institute for Transport Studies, University of Leeds.

Professor David M Newbery, Department of Applied Economics, University of Cambridge.

Philip O'Donnell, Assistant Director, Office of Passenger Rail Franchising.

Dr John Preston, Director, Transport Studies Unit, University of Oxford.

Trevor Soames, Partner, Norton Rose.

David Starkie, Director, Economics-Plus Limited.

Foreword

JOHN PRESTON
Series Editor

This volume represents the second in the re-launched Oxford Studies in Transport Series. We are grateful to Valerie Rose of Ashgate for her continued support. This volume continues the regulatory theme established by the first volume in the series[1]. It is based on a seminar held in Oxford in September 1998, organised largely by Helen Lawton Smith, but with some assistance from David Starkie and myself. I would particularly like to thank Ann Heath for undertaking the tortuous process of converting various word processed files into the standard required by Ashgate, and Angel Trains for providing the appropriate financial support. I would like to also thank the contributors and the publishers for their patience whilst this somewhat protracted process was completed.

[1] Van de Velde, D. (Ed.) (1999) *Changing Trains. Railway Reform and the Role of Competition: The Experience of Six Countries.* Oxford Studies in Transport. Ashgate, Aldershot.

1 Introductory Overview

HELEN LAWTON SMITH, JOHN PRESTON AND
DAVID STARKIE

Background

The inspiration for this book was provided by the Regulatory Policy
Research Centre's seminar on *Integrated Transport Policy: Implications
for Regulation and Competition* held at Hertford College, University of
Oxford on the 24th and 25th September 1998. This book is based on the
papers presented at that seminar and subsequent reflections. The 1998
seminar was a follow-up to the seminar on *Privatisation and Deregulation
in Transport* held at Hertford College, University of Oxford, on the 2nd to
4th July 1997. The proceedings of this seminar have also been published
(Bradshaw and Lawton Smith, 1999). Both seminars were part of a series
run by the Economic and Social Research Council's Regulatory Policy
Seminar Series (award numbers R45126440395 and R45126472097). The
1998 seminar was also part funded by Angel Train Contracts. We are
grateful for the support of ESRC and Angel Trains, without whom this
book would not have been produced.

The policy background was provided by the publication in July 1998 of
the Transport White Paper 'A New Deal for Transport: Better for
Everyone'. This White Paper represented a significant shift in the UK
Transport Policy, albeit one that was initiated by the Great Transport
Debate and the 1996 Green Paper, 'Transport: The Way Forward'. At the
time of the seminar, there was a degree of excitement over the new
developments in transport policy. Despite the subsequent publication of a
whole raft of daughter documents and other related materials, this
excitement has dissipated, mainly due to the lack of primary legislation.
The new focus of transport policy is to be integration, both within and
between modes, with environmental policy, with land-use planning and
with education, health and wealth creation policies. This would seem to
imply a more interventionist approach by Government than had hitherto
been the case (at least since 1979). The underlying hypothesis of this
seminar was that this would create tensions with pro-competition policy
(including the 1998 Competition Act) and have significant implications for

regulation. These tensions are also apparent in European transport policy, where following the belated publication of a White Paper on a Common Transport Policy in 1992 (35 years after the founding treaty) there has been a major push towards liberalisation (see Preston, 1999).

Structure of the Book

Given the above background, the structure of the book is as follows. Chapter Two, by Ken Button, provides the second part of the introduction by tracing the evolution of transport policy, particularly with respect to the United Kingdom but also with respect to developments in Europe and globally (particularly in the United States).

The second part of the book focuses on the pricing of infrastructure. In Chapter Three, Chris Bolt compares Railtrack's charging regime with that proposed for rail infrastructure by the European Commission as part of the Fair and Efficient Pricing Initiative. In Chapter Four, David Newbery examines the financing and pricing of road infrastructure. In Chapter Five, David Starkie provides a commentary on the previous two papers, comparing and contrasting the approaches adopted in the rail and road sectors.

The third part of the book examines the effectiveness of competition. John Preston examines the impact of competition initiated by the 1993 Railways Act in Chapter Six, whilst John Dodgson in Chapter Seven examines the impact of competition in the local bus market initiated by the 1985 Transport Act. Chris Nash provides a commentary on these two papers in Chapter Eight.

The fourth, and final, part of the book examines the adequacy of the regulatory framework in the United Kingdom. In Chapter Nine, Bill Bradshaw examines the range of regulations applicable to the rail and road industries and highlights the important issue of enforcement. In Chapter Ten, Mike Hoban examines the implications of the 1998 Competition Act, essentially an application of Articles 85 and 86 of the 1957 Treaty of Rome, now somewhat confusingly re-numbered Articles 81 and 82 as part of the 1997 Treaty of Amsterdam. This legislation will have important implications for many of the policies proposed in the 1998 White Paper, particularly concerning Quality Partnerships in the bus industry. In Chapter Eleven, Philip O'Donnell provides a commentary on Bradshaw and Hoban's paper, raising amongst other things, the role of merit tests, the free rider problem in quality upgrades and the importance of dynamics. In Chapter Twelve, Trevor Soames examines European policy towards essential facilities and reviews important cases concerning Frankfurt and Paris airports, SABENA's computer reservation system and the ports of Holyhead, Rodby, Elsinore and Roscoff. In Chapter Thirteen, Melanie Farquharson examines essential facilities case law in Britain, particularly with respect to European Night Services (rail) and bus stations in Newport (Isle of Wight) and Chatham (Kent).

From the above, it should be clear that the emphasis of this book is on land-based passenger transport, particularly rail and road. The main exception is Chapter Twelve which necessarily concentrates on the air and sea sectors. Furthermore, the emphasis has been particularly placed on passenger rail and bus. It must be borne in mind that in some respects these are declining industries. In 1952, rail and bus accounted for 60% of passenger travel by mechanised modes in Great Britain. By 1997, bus and rail's market share had reduced to 12% (DETR, 1998). Any policy prescriptions recommended for these modes may not be applicable for growth sectors of the transport industry such as air, car and road freight.

Overview

In the remainder of this introductory chapter we highlight some common themes that emerge throughout the book.

The first is that of policy cycles, raised by Ken Button in Chapter Two and Mike Hoban in Chapter Ten. Both authors compare the 1998 White Paper with the 1966 White Paper Transport Policy, which was also concerned with integration. Only time will tell whether Ken Button's assessment is correct - namely that the 1966 White Paper will prove more important than its 1998 counterpart. What is clear is that these policy cycles are linked with the regulatory and ownership cycles that have been identified by others (for example, Needham, 1983). In terms of regulation, bus was regulated by the 1930 Road Traffic Act and largely deregulated by the 1985 Transport Act. Some re-regulation in the near future seems possible. Railway regulation was introduced progressively from the 1844 Railways Act and also partially deregulated progressively, with the 1962 Transport Act being particularly important. The 1993 Railways Act effectively saw the re-regulation of the passenger rail industry, although the rail freight industry was further deregulated. In terms of ownership, the bus industry was taken into public ownership progressively with the 1933

London Passenger Transport Act being an important turning point. The 1985 Transport Act initiated the return to private ownership. Railway nationalisation was achieved by the 1947 Transport Act and the return to private ownership by the 1993 Railways Act. This potted history indicates that regulatory and ownership cycles are only partially synchronised both within modes and between modes. There is also an important contrast between a big bang approach (bus regulation, rail ownership) and a more gradual approach to implementation (bus ownership, rail regulation). Button (1998) finds similar contrasts between the Anglo-Saxon and European approaches to regulatory reform. An interesting question is why do such cycles occur? (On this, see also Starkie, 1984). Partly it is, of course, due to the whims of the ballot box but it does appear that the dynamics of regulatory processes lead to alternating concerns regarding market and regulatory failure. Robinson (1999) points out that many regulatory analyses involve an unfair comparison of imperfect competition with perfect government, that there are strong incentives for governments to over-expand their regulatory activities (not least because 'external' compliance costs may be 50 times 'internal' administrative costs) and that there are dangers of regulation crowding out alternative competitive solutions. Although Robinson concludes in favour of market-based solutions, except where regulatory activity is pro-competitive, his analysis highlights some of the key elements in regulatory dynamics.

The second issue, which is essentially the crux of part two of the book, is the on-going debate concerning whether transport pricing should be based on long or short-run marginal costs (or as Chris Bolt suggests, simple variants thereof). The text-book answer is that, given optimal investment and constant returns to scale, production should be at a level where short and long-run marginal cost coincide. Clearly we do not live in such a first best world and hence we need to determine to what extent investment diverges from the optimal and to what extent there are increasing or decreasing returns to scale. Starkie, in Chapter Five, contends that railways exhibit broadly constant returns to scale, although they often exhibit increasing returns to density. By contrast, he believes that roads, and particularly urban roads and road widenings, exhibit decreasing returns to scale. More empircal work on the scale economies in the road and rail industries is required, building on the work of Kraus (1981) and Caves et al. (1985), amongst others. Attention also needs to be paid to the dynamics of the price: investment relationship. This is a particular weakness of the current road pricing debate, as identified by David Newbery. Although there are obviously difficult issues here, one cannot help but feel that the experience from other sectors, particularly the energy and telecommunications sectors, has been somewhat underutilised in transport (see, for example, the texts by Rees, 1984, and Brown and Sibley, 1986).

A third issue relates to the debates concerning the efficacy of competition in the market and competition for the market. This is, to some extent, the sub-text of part three of the book, but is also touched on by Hoban in Chapter Ten who states that:

> Competition in the market is preferable to competition for the market.

Hoban argues that competition for the market stifles innovation, leads to extra administrative costs, tends to lead to over-investment and reduced service quality. The evidence in Chapters Six and Seven is more mixed. Preston judges that competition for the market has been successful in terms of franchising the passenger rail market but speculates that competiton in the market will be less successful. Dodgson provides statistics that indicate that tendering of buses in London has been more successful than competition in the market for buses outside London, but concludes that this was mainly due to favourable external circumstances in London (including a rising population and falling car ownership). It is interesting to speculate about the scope for contracting out the planning function as well as the operations function of public transport industries and whether that would reduce some of Hoban's concerns. Bradshaw, in Chapter Nine, presumes in favour of high quality competiton but against low quality competition. Dodgson introduces the concept of horizontal product differentiation (consumers will choose different products at the same price) and vertical product differentiation (consumers will only choose the highest quality product at the same price). Using this terminology, Bradshaw seems to favour vertical product differentiation (at least when that involves the introduction of higher quality services) over horizontal product differentiation (at least when that involves additional low quality services). The problem here, as Dodgson identifies, is that there may be limited scope for quality competition within conventional public transport industries, although one would expect this statement to be more valid for bus than rail given the relative homogeneity of bus users. Some further theoretical and empirical work may be required here. Nash, in Chapter Eight, reflects on the appropriateness of contestability theory. This theory has some attractions in that competitive outcomes in terms of prices and outputs may occur without actual competition (and the risks that such competition could be wasteful). Sadly, contestability theory did not provide the uprising in industrial economics its proponents hoped for. The bus industry has not proved to be perfectly contestable due largely to a combination of short reaction periods and reputation effects. Although the rail industry may have longer reaction periods, it may also have more substantial barriers to entry related to staff training and safety regulation. Paradoxically, despite some exit barriers, competiton for the market may be more contestable than competition in the market. Nash also points out that there are strong first best arguments (the Mohring effect) and second best arguments (the decongestion effect) to subsidise urban public transport which is problematic for a contestable market structure. Furthermore, where there are budget constraints, there may be a case for cross subsidy which causes further problems for a contestable structure. Further competition models might also be considered including intrafirm quality competition (e.g. vehicle quality - see Kain and Starkie, 1998), intramodal route competition, inter modal competition, upstream and downstream product market competition and competition by emulation (yardstick competition) (see, for example, Van de Velde et al., 1998).

A fourth issue, and one related to the third, is the debate concerning the relative importance of productive, allocative and dynamic efficiency. As flagged by Ken Button in Chapter Two, the concern of British transport policy in the 1980s was with productive efficiency with a rule of thumb that cost reductions of 30% could be achieved through regulatory and organisational reform. There was also concern about the lack of dynamic efficiency in public transport industries, reflected by a poor record in terms of innovation. In the late 1990s concern has perhaps re-centred on allocative efficiency, as it appears that the bus industry outside London has tended to provide more service at higher fares than demanded by customers, whilst the rail industry has failed to provide the quality that seems to be demanded by its customers. There has, however, been surprisingly little attempt to quantify the efficiency effects of British regulatory reforms in public transport. Exceptions include Mackie et al. (1995), Harris and Godward (1997) and White (1998), although the calculations are not uncontentious.

A fifth issue is that of transitional costs and transaction costs. The re-organisation of the British railway industry into a market based system based on 70,000 negotiated contracts and over 100 new undertakings has clearly involved substantial transitional and on-going transaction costs. As Philip O'Donnell stresses in Chapter Eleven, these costs need to be quantified and compared with the efficiency gains of market testing. Is the current regime an improvement over the previous regime which was essentially an administered system based on command and control? Has information technology reduced the transitional and transaction costs?

A sixth issue relates to the most appropriate types of regulation. Bradshaw provides a functional classification by examining regulation with respect to investment, service quality, fares, safety and competition and notes the different degrees of regulation (and enforcement) in the road and rail sectors. O'Donnell makes a useful distinction in Britain between contract-based systems, for competition for the market, and licence-based systems, for competition in the market. The latter is also used as the basis for the more traditional regulated systems found in much of continental Europe (see, for example, Van de Velde, 1999).

A seventh issue is that of vertical separation. This was much less of an issue for the bus and coach industry, despite the evidence from express coach deregulation that terminals were an important barrier to entry. In the event, the 1985 Transport Act separated the ownership of National Express and London Victoria Coach Station but assumed existing competition policy could deal with other cases of what Trevor Soames calls leveraging, which as Melanie Farquharson reports is effectively what happened. Vertical separation is more contentious in the rail industry with both Preston and Bradshaw critical of arrangements in this respect. Disadvantages include strategic concerns over investment levels, rail transport and land-use integration and safety, tactical concerns about the scheduling of maintenance, the attribution of blame and the provision of information, and operational concerns regarding the scheduling of delays. Advantages include the greater degree of specialisation, the emerging transparency of costs and the greater impetus vertical separation gives to open access, not least because, as Melanie Farquharson points out, it circumvents aspects of the essential facilities problem. Yet again more theoretical and empirical evidence is required before the balance of advantage can be determined.

The eighth issue concerns the Europeanisation of British competition policy. In essence, Chapter I of the 1998 Competition Act is a re-working of the (old) article 85 on cooperative agreements and Chapter II is a re-working of article 86 on dominance. Although the new legislation is likely to be an improvement on the previous rather ineffective legislation, it seems that European policy may be rather simplistic and has failed to incorporate some of the subtleties of American anti-trust economies (see, for example, Tye, 1992). The two positive and two negative conditions that are required to be met if a cooperative agreement is to be permitted seem to be rather all embracing and, with respect to the Government's aim to achieve an integrated transport policy, as Hoban points out, leads to a concern

> that the ambition to facilitate modal shift initiates the prospect of on-road competition to the extent that entry barriers are set too high; or low cost potential entrants are banned, or restrictions are placed as the services they provide, leading ultimately to higher prices to customers.

With respect to predation, application of the AKZO test (where price is less than average variable cost then it is judged predatory, when it is between average variable and average total cost it is possibly predatory) there may be dangers of asymmetric regulation with entrants getting more protection than incumbents. Some competition may be caused by erroneous entry. However, the AKZO test is merely a European application of the Areeda-Turner rule used in US anti-trust, which has been argued by some to be too permissive, allowing predators to escape too easily (Easley et al., 1985). Further empirical evidence is required but it seems that either more sophisticated 'bright lines' are required or that a case by case approach should be adopted (Dodgson et al., 1993).

The last issue is that of materiality. Little attention has been paid to the issue of when is market failure material enough to justify intervention. It may be that many of the economic gains of regulation are relatively modest, particularly when set alongside the risks of reducing dynamic efficiency. Similarly, when is a market substantial enough to warrant investigation? Both British experience, where relatively small bus markets have been investigated, and that of Europe, where a whole host of small seaports have been investigated, suggest such markets may be relatively small in geographic scale and product definition. Also, when does dominance become an issue and when do profits run the risk of being identified as excessive? Indeed, are profits per se an indicator of market power, and hence of relevance to pro-competition authorities? Both Soames and Farquharson hint that such materiality tests may be influenced by individual personalities. This may be a particular problem for the UK where regulation of utilities has been complicated by this factor.

What the above suggests is that there is a whole host of regulatory issues which remain unresolved. However, the past twenty years of experimentation in the transport sector in Britain has led to a better understanding of many issues, albeit with some new issues also emerging. There is an increasing realisation that the policy menu is more extensive than previously thought. Furthermore, the sequencing of policy has emerged as a key issue. For example road pricing may need to be preceded rather than followed by public transport improvements (including better regulation). It is to be hoped that Britain is moving towards a better transport policy and a better regulatory policy, although not necessarily in that order. As always, history will be the judge.

References

Bradshaw, W. and Lawton Smith, H. (2000) *Privatization and Deregulation of Transport.* Macmillan, London.

Brown, S.J. and Sibley, D.S. (1986) *The Theory of Public Utility Pricing.* Cambridge University Press, Cambridge.

Button, K. (1998) The Good, the Bad and the Forgettable - or Lessons the US can Learn from European Transport Policy. *Journal of Transport Geography*, 6, 4, 285-294.

Caves, D.W., Christensen, L.R., Tretheway, M.R. and Windle, R.J. (1985) Network Effects and the Measurement of Returns to Scale and Density for US Railroads. In Daughety, A.F. (Ed) *Analytical Studies in Transport Economics.* Cambridge University Press, Cambridge.

DETR (1998) *Transport Statistics Great Britain. 1998 Edition.* The Stationery Office, London.

Dodgson, J.S., Katsoulacos, Y. and Newton, C.R. (1993) *Application of the Economic Modelling Approach to the Investigation of Predation*, 27, 2, 153-170.

Easley, D., Masson, R.T. and Reynolds, R.J. (1985) Praying for Time. *Journal of Industrial Economics*, 33, 445-460.

Harris, N.G. and Godward, E.W. (1997) *The Privatisation of British Rail.* The Railway Consultancy Press, London.

Kain, P. and Starkie, D. (1998) *Overcrowding on Commuter Trains: An Economic Solution.* Economics-Plus Perspectives, June.

Kraus, M. (1981) Indivisibilities, Economies of Scale and Optimal Subsidy Policy for Freeways. *Land Economics*, 57, 115-121.

Mackie, P., Preston, J. and Nash, C. (1995) Bus Deregulation: Ten Years On. *Transport Reviews*, 15, 3, 229-251.

Needham, R. (1983) *The Economics and Politics of Regulation: A Behavioural Approach.* Little, Brown and Company, Boston.

Preston, J. (1999)   The Future for Competition and Ownership in European Transport Industries.   *Proceedings of the European Transport Conference,* University of Cambridge.

Rees, R. (1984)   *Public Enterprise Economics* (Second Edition).   Philip Allan, Deddington.

Robinson, C. (1999) The Perils of Regulation. *The Utilities Journal,* 2, 2, 34-36.

Starkie, D. (1984) Policy Changes, Configurations and Catastrophes. *Policy and Politics,* 12, 1, 71-84.

Tye, W.B. (1992) Market Imperfections, Equity and Efficiency in Anti Trust. *The Anti Trust Bulletin,* 37, 1, 1-34.

Van de Velde, D. (1999) Organisational Forms and Entrepreneurship in Public Transport. Part 1: Classifying Organisational Forms. *Transport Policy,* 6, 3, 147-158.

Van de Velde, D.M., Mizutani, F., Preston, J. and Hulten, S. (1998)   Railway Reform and Entrepreneurship: A Tale of Trhee Countries. *Proceedings of the European Transport Conference.* PTRC London.

White, P. (1998) Financial Outcomes of Rail Privatisation in Britain. *Proceedings of the European Transport Conference,* Seminar G, Rail, PTRC, London.

2 The Evolution of UK Transport Policy

KENNETH BUTTON

Introduction

Some years ago I made the statement, '...Government finds the transport sector extremely difficult to handle; this is perhaps most clearly seen if one reflects on the fact that *major* pieces of transport legislation have appeared on the statute books every seven or eight years during the past half-century (i.e. 1930, 1933, 1947, 1953, 1962, 1974, 1980)' (Button, 1982). The passing of the 1985 Transport Act and the Railway Act eight years later provide on-going confirmation of this trend. It is apparently time for another burst of activity.

Over the years, successive UK Governments have also produced a number of important policy statements setting down how they intend developing the nation's transport infrastructure and controlling its use. The current White Paper (UK Department of Environment and Transport, 1998) should be seen as part of that policy continuum; indeed, the emergence of *A New Deal for Transport* offers reassuring confirmation of the world's stability.

The policy set out in the White Paper seeks to develop a strategy that will, 'create a better, more integrated transport system to tackle the problems of congestion and pollution' (p.3)[1]. Not a modest aim, but equally one that few would have any qualms about supporting in principle. The issues, as nearly always, with official transport policy statements are not with the broad objectives but rather with the detail and with the means of implementation.

This paper is not so concerned about offering a full and rigorous critique of *A New Deal for Transport*; that would take up too much space and would also require the presentation, either implicitly or explicitly, of counter proposals that are beyond our scope. Rather, the paper is concerned with looking at where the policy proposals sit in this continuum of policy statements over the years. It aims to highlight, in particular, how the Labour Party's thinking on transport issues has changed since the 1960s

11

when it published its last really important policy statement on the subject. In doing this there is inevitably some discussion of the merits of current policy but this is largely secondary to the main thrust of the paper.

Initially, the proposals are set out in their general historical context. Although there have been some notable exceptions, 'new' policy seldom represents a sea-change in thinking but fits in with longer term trends and evolutionary developments (Starkie, 1984). It is wrong, therefore, to consider it in total isolation. This paper proceeds rather selectively by considering the objectives of the proposals, the constraints that confront those developing the policy and, the policy instruments that have been favoured. This provides a prelude to a comparison with what the Labour Party was attempting to do thirty years ago.

The Context

Transport policy goes in cycles (Button and Gillingwater, 1986). These cycles extend well outside of the UK. The latter part of the last century and the early decades of this saw an emphasis on controlling monopoly power and witnessed the introduction of economic regulations over price and market entry across a wide range of transport industries. The inter-World Wars period witnessed an up-surge of regulatory controls over transport particularly with the passing of the Road Traffic Act in 1930 and the Road and Rail Traffic Act in 1933. The immediate post-1945 period saw consolidation of the regulatory structure through an expansion of public ownership of the railways, airlines and much of the road haulage industry (notably under the Transport Act of 1947). The 1960s saw a move to controlled competition that allowed a greater play of market forces but largely within a tightly restricted regulatory environment (e.g. the 1968 Transport Act).

The 1980s might generally be called 'The Age of Regulatory Reform' (Button and Swann, 1989). Major regime shifts took place in virtually all the traditionally market driven economies of the Organisation for Economic Cooperation and Development (see OECD, 1994) but the title is also apposite in the wake of events in Central and Eastern Europe. Transport has been at the forefront of this movement. Internationally, regulated transport industries were deregulated or the regulatory regime liberalised (Button and Keeler, 1993). Privatisation of infrastructure and operations occurred. Other forms of economic intervention, such as subsidies, tended to decline. The UK was at the forefront of these reforms[2] with, for example, long distance and local bus services largely deregulated and privatised by the Transport Acts of 1980 and 1985; and the railways, airports, some sea ports and airlines privatised although often under a regulated structure[3]. The European Union, after staggering down a number of blind alleys, began to produce a coherent Common Transport Policy for the Union that for a large part favoured minimal regulation and large scale private sector involvement.

The latter part of the 1980s and the 1990s have seen an increasing concern with the issue of integrated transport within the wider framework of sustainable development. The concept, popularised with the publication of the Brundtland report (World Commission on Environment and Development, 1987), seeks to highlight the need to provide a legacy of resources for future generations. While holistic in its orientation, the notion of sustainable development necessitates sectorial implementation. Efforts at achieving this have been thwarted by conceptual difficulties as well as by more mundane matters of implementation. The European Union has sought to devise some common ground for moving toward a more sustainable transport environment in Europe but this is only slowly moving from the embryonic stage (Commission of the European Communities, 1992).

This mainly, environmental concern has recently coexisted with a more traditional and narrow transport concern. This is traffic congestion. The basic economic costs of pursuing previous strategies, that in themselves did little to limit transport but rather focused on accommodating and managing, have become unacceptable. The rate of increase in transport use has led to an almost Malthusian situation in many people's eyes that while transport is growing rapidly the infrastructure needed to accommodate the growth simply cannot be made available at the same pace even if it were environmentally benign[4]. General traffic restraint has become a key concept in transport policy[5].

What causes these policy fluctuations? These cycles are partly a function of prevailing perceptions as to how the transport system operates and partly a reflection of the wider policy objectives of the day. The recent upsurge in concern about the implications of greenhouse gas emission from automobiles, for example, reflects new scientific findings. Policy shifts are also coloured by both the nature of the external environment in which transport operates and by the experiences of previous policy initiatives[6]. The coming together of economic and political ties within Europe has been influential in the way international transport policy is now treated. The outcomes of the deregulation of buses in the 1980s has highlighted some of the market failures that require re-examination in this sector. There are also technical factors to consider. The rapid developments in information technology provide the possibility for the adoption of a range of new policy instruments (e.g. regarding electronic road pricing and electronic guidance systems). They also affect the way that transport is itself being supplied (e.g. through its use in logistics management and real time information systems for travellers).

It is a fine judgement for a Government as to when to initiate a new major transport policy initiative. When is it appropriate to move to a new phase in the policy cycle? There are natural, all round pressures for an incoming administration to want to create a clean slate and to make its mark on policy. Transport is an obvious candidate for this. It is a highly visible sector that few people are ever satisfied with; perhaps they never can be. The sector is both concentrated (e.g. in terms of infrastructure) and diffuse (e.g. in terms of car users) which leads naturally for calls of better organisation and integration. People are always encountering problems (e.g. incidents leading to congestion) but seldom see themselves as a problem. There are spectacular system failures (e.g. air crashes). There is a natural inclination to say that there must be a better way. The real issue, though, is whether the Government has anything really new to say.

In a sense perhaps the notion of a 'New Deal' is very appropriate for transport policy at this time. It may be, of course, that the Secretary of State has a hidden personal agenda and sees himself leaving a legacy every bit as enduring as President Roosevelt's in the US. More importantly, and certainly more realistically, the title of the White Paper may be a subconscious reflection of the key underlying message of the Roosevelt era, namely that many problems need a change in attitudes as much as government policy[7]. If one reads the document it is clearly evolutionary not revolutionary but what it does do is very clearly spell out what the important transport issues of the day are and how there is a need to fight today's battles not yesterday's; the larger arguments over privatisation and regulation are really largely settled. This need to move on was inherent in much of the discussion during the later part of the previous Conservative Government's stewardship, but concern also still seemed to linger on tidying-up the ends of the 'Age of Regulatory Reform'.

The Objectives

The objective of the policy set in the White Paper is to, 'extend choice in transport and secure mobility in a way that supports sustainable development' (p.10). These are fairly unexceptional goals and are not out of line with those intimated at in the Conservatives' Green Paper, *Transport the Way Forward*. Here it was also recognised that society sought choice in the way that transport was supplied but were concerned about the environmental intrusion that transport creates.

What both these set of objectives do is conflict with the core of policy in the 1980s and early 1990s where objectives were seen much more in terms of enhancing the economic efficiency of the UK's transport system. It also sought to redefine the important elements of efficiency with its focus shifting away from attention being over allocative and scale concerns to an orientation on technical and dynamic efficiency. This involved extensive deregulation and private sector involvement in supply. The new White Paper is not silent on the question of efficiency, either in its older form or its more recent manifestation, but it is treated in a somewhat different manner.

The Conservatives' concern was mainly in terms of reducing the degree of intervention failure that existed and to contain the extent of regulatory capture in the transport system. Greater recourse to markets was seen as the primary solution with more direct forms of regulation when market failure or social considerations were felt important. The current document is implicitly more concerned with the market failures that have emerged in this deregulated climate - e.g. in the context of through-ticketing (p.48), inadequate taxi services (p.53) and empty running of freight vehicles (p.71). Massive re-regulation is not an objective but a marginal tidying up process is effectively being proposed.

While it is never fully debated, the notion of integration is taken as a primary objective. This integration involves a wide range of dimensions both internal to transport and with respect to the interface between transport and other activities. Implementation involves a variety of policy instruments and some institutional reforms.

Many would feel that integration is a means to an end rather than an end in itself. Setting this aside, more importantly, integration can take a number of different forms and there are many ways integration can be brought about. In some instance in the paper it would seem that there is clearly no strong view on the nature of the integration to be adopted and there is frequent mention of the need for further consultation. This consultation may be between Government and the private sector or between different branches of Government. This need for additional consultation does leave some gaps that need filling for the paper to be fully definitive.

Integration can also take many forms - the Labour administration of the 1960s tried to integrate transport supply through what amounted to 'controlled competition' within an urban planning structure[8], while the post-1979 Conservative Government's market based philosophy was also a form of integration.   In the interim, the Labour Party of the mid-1970s advocated 'guided choice' for the transport market[9]. It is not altogether clear where this document stands in the spectrum of possible integration doctrines.   Its omission of any extensive discussion of privatisation indicates that markets are an important tool, the discussion of pricing and taxation instruments indicates a preference for integrating some areas of policy through economic instruments (other than markets) while at other times the increased level of public funding and physical control measures points to a more regulatory attitude. The approach might, given there is apparently no clear underlying concept, perhaps, best be called 'pragmatic integration'.

The Constraints

Official policy documents generally spend considerable time outlining objectives and in putting forward preferred solutions. They spend much less time looking at constraints and costs and seldom mention anything about fall-back positions if the primary addenda prove to be a failure. They are, after all essentially political statements.

One of the biggest constraints on initiating any policy is its public acceptance.   The Conservative Government learned this with the Community Charge but more parochially local authorities, such as Cambridge, met opposition to local road user charges. One of the problems in transport policy, to reiterate an earlier point, is that while the public want less congestion and less pollution it is always the other people that must change their attitudes and life-styles. Education may help, but the evidence is that people are actually aware of the problems but they still refuse to change their behaviour accordingly. While the notion of sticks is not new in transport policy, they may perhaps eventually have to become more pronounced. The White Paper accepts this but is rather limp in terms of pushing for radical change. One can understand this.

The demand for transport services is derived from the demands of individuals and firms. These final demand patterns, while certainly not rigid, change slowly. Recent years have seen developments in such things as the composition of UK industry and patterns of land-use that inevitably limit the options open to transport policy makers. The growth in importance of service sector activities combined with the decline in heavy manufacturing industry makes service quality (e.g. speed, frequency and reliability) an important component of any transport system. New production techniques embodying notions such as just-in-time management are a natural response to this. In terms of passenger travel, a maturing population, sub-urbanisation, increased leisure time and disposable income and changes in job activity rates, mean people are seeking different attributes from transport compared to twenty or thirty years ago. These demands are not easily modified in the short term.

The UK is also no longer entirely free to set its own transport policy agenda. Historically, because of the geography of the country and its extensive trading activities, there have always been constraints over the way that shipping and aviation policy could be designed. Membership of the European Union poses additional constraints[10]. In many ways the UK is fortunate. In general, the transport policy of the EU has moved toward the Anglo-Saxon doctrine that has been the basis of UK transport policy since the late 1960s. The need for efficient transport to meet the needs of the Single European Market initiative, combined with the realisation that on-going subsidies in Europe were neither politically nor economically sustainable in the longer term, led to market liberalisation and privatisation (Button, 1992). It would be very difficult given the political structure of the EU and the impetus for what is often called 'deregulation', for the Labour Government to make any major moves in the opposite direction.

The UK's involvement in international transport is also growing outside of the EU area as long distance tourism grows in popularity and as trade becomes more global. Many aspects of policy reform here cannot be undertaken unilaterally. This is the case of international air transport where bilateral air service agreements are relevant (p.101) and extends to some areas of environmental policy. Lack of independence, or the existence of wider agreements, inevitably bind what can be done within parts of the UK transport sector.

The Policy Instruments

The Conservative's transport policy shifted the emphasis still further from command-and-control regulation and towards more market based policy instruments. Privatisation was a clear manifestation of this. The current policy document retains this focus on the efficiency of markets with its discussions of road pricing (p.115-6), working-place parking charges (p.117), tax protection for bus operators (p.122) and vehicle taxation for environmental purposes (p.121)[11].

The difference is that these fiscal tools are taken as devices to achieve particular targets whereas the notion of markets and privatisation has much more to do with allowing optimal supply and demand conditions to evolve internally. The current Government has less opportunity to use genuine markets because most of the more obvious applications have already been realised. In many of the areas where the White Paper advocates the use of economic instruments it is difficult to see how markets could be put into place anyway. Traffic congestion is a 'club good' problem and private provision of roads with charging could generate market conditions but containing greenhouse gas emissions poses near insurmountable problems of property rights allocation. Markets do not function without such allocations. Since markets are not always possible to initiate, even in a workably competitive form, use of fiscal instruments does provide a second best alternative.

One of the possible limitations of the usefulness of the policy instrument portfolio set out in the White Paper is, however, the autonomy left to local government. Integration of policy should embrace spatial integration but policies such as road pricing are to left to the discretion of local governments; essentially an enabling doctrine (p.115). The problem with this is that for road pricing to be effective it relies upon a consistent application. In economic terms road pricing is only a first best strategy if all prices are set at marginal cost. The incentive to use road pricing, and its effective application in those cases where it is deployed, is much greater if local authorities do not fear commercial desertion to adjacent areas.

Attracting people to non-motorised modes of transport for health as well as congestion reasons marks an appreciation that the range of transport modes is much larger than traditionally thought (although the absence of any reference to the flexible roller blade mode is a sad omission, especially since it avoids the parking and security problems bicycles entail)[12]. While there are eminently good reasons for ensuring pedestrians and cyclists enjoy safe and good infrastructure, the case for assuming that cycling in particular offers a substitute for motorised trips is weak. Indeed, the evidence offered in the paper alludes to the number of trips now made by bicycle in several cities not the transference from the private car[13].

A Comparison

The last major initiative on transport from the Labour Party was that of the 1960s; the Government of the late 1970s being largely preoccupied with other issues. Rather than contrast current Labour Party thinking with that of the previous Conservative administration, from which the current strategy may be seen in many ways evolutionary rather than revolutionary, it is perhaps interesting to see how the current Labour administration differs in its approach to transport matters from the Wilson Government of the 1960s.

There are some similarities and differences in the background against which the policies emerged. The 1960s was a major watershed in terms of transport policy making in the UK and laid the foundations for the broad policy framework that effectively endured until the Thatcher reforms of the 1980s. Policy was developed, however, at a time of near crisis in both the railways and bus transport sectors and followed a period of extended macro-economic crises. The railways had been accumulating significant losses while the regime of cross-subsidisation for bus services that had been initiated in the 1930s was failing to cover the systems' costs. There were certainly long term, structural issues involved in the policy of the 1960s but the policy was also coloured by short term practicalities. The late 1990s, in contrast, sees the UK's transport system being stretched as congestion grows and under pressure to improve its environmental performance, but there are no actual crises to manage. The UK macroeconomy has enjoyed a period of extended growth and reasonable stability, providing a resource and confidence base that can be drawn upon.

The fact that the 1960s Labour administration had been in power for some time when it introduced its major transport initiative also meant that it had greater flexibility to act. The current document frequently points to the need change policy later because existing contracts have time to run; this applies, for example to rail fares (p.95) and rail service levels (p.96).

Another important difference was that the Labour Government of the 1960s came to office with much less understanding of the transport sector than its current counterpart. Transport data was sparser and analysis less complete at that time. The Labour administration was responsible for drawing a large number of experts into the Ministry of Transport and related agencies to develop the traffic forecasting models and policy analysis tools that, suitably up-dated, form the basis of much transport analysis today. The 1960s saw a massive outpouring of studies[14] and consultation documents prior to the eventual emergence of the 1968

Transport Act[15]. The policy that has emerged 30 years later has appeared somewhat more rapidly and with less public consultation[16]. This is understandable; a certain amount of basic knowledge is now available and accepted. One might anticipate, therefore, that current policy recommendations would be more authoritative in their tone. In fact, as indicated above, this is not so.

In terms of documentation and the marketing of policy clearly much has been learned over the years. The White and novel Green (soon becoming Pink) papers of the 1960s were simply text, there was the odd table or black and white graph for light relief but certainly no colour pictures, and the jargon was somewhat more technical, often almost legalistic, than we find in *A New Deal for Transport*. Public consultation and policy interaction seemed aimed at a more narrow audience in those days. While much of the current paper is well put together, one must wonder whether it is worthwhile trying to attract the infant school audience by relating 'Kate's Story', 'Sue's Story' and 'Joe's Story'.

The overall objectives of the 1968 and 1998 policies, however, emerge as very much the same. Munby (1968, p.136) summed-up the 1968's approach as an attempt, 'to grapple with the overall market situation and to use all means of policy'. In 1998 we find that the claimed coverage is multimodal and concerned with all aspects of policy (p.13). Both positions claim to be reflective of the need for a change in the *status quo*.

Freeing the market from excessive Government intrusion was a key element of the 1960s reforms. Road freight transport was *de facto* freed from quantity controls and the railways were given greater commercial freedom. The reforms have continued since that time and the current White Paper, largely through what it does not say, does not substantially buck that trend. There are hints of a lessening of momentum - e.g. the decision not to privatise the trust ports (p.102) and to provide local powers to ensure bus operators participate in multiple ticketing (p.48) - but nothing by way of a major reversal of the deregulation and privatisation that has already been carried through. Indeed, the decision to partially privatise the National Air Traffic Services (NATS) (p.101) represents a progressive move to more private sector participation in transport infrastructure provision.

Traffic congestion was a major focus of attention in the 1968 legislation but despite this, and efforts of successive governments, the situation has not improved. Urban traffic congestion is now complemented by increasing congestion of inter-urban networks. The policy makers of the 1960s, while paying lip service to the possible future use of fiscal measures such as road pricing, relied upon a combination of physical restraint, public transport alternatives and infrastructure construction in their holistic approach. The current policy position is that, while the conventional instruments of land-use planning and public transport support, albeit in revised form, are necessary and important, economic instruments need to be used more widely. This entails such things as allowing '...local authorities to charge road users so as to reduce congestion...' (p.115), introducing on trunk roads and motorways pilot charging schemes (p.116) and allowing 'local authorities to levy a new parking charge on workplace parking' (p.117).

The environmental debate was an important element in policy formulation thirty years ago and remains so today. The only difference is that the definition of the 'environment' has widened somewhat. In 1968 the issue was mainly about making the urban environment tolerable under the weight of increasing volumes of noisy and polluting cars and lorries[17]. Policies of land-use planning (following the Buchanan doctrine) and administratively integrated public transport systems to attract people from their cars were adopted. The policy initiated the idea that there should be, 'unified planning of transport under a single Transport Authority' (UK Ministry of Transport, 1967).

Today many of the local environmental problems that concerned people in the 1960s have been reduced. Lead is now largely removed from petrol, due to differential taxation and the banning of some grades of leaded fuel, and NO_x emissions have been cut through the use of catalytic converters. Local pollution problems associated with aromatics remain and may have worsened. The key new issue is that of greenhouse gases. This is a problem that cannot simply be tackled by engineering solutions (although greater fuel economy of engines helps considerably) but requires an actual reduction in the overall use of carbon fuels. Again, economic instruments are viewed as playing a larger role than in the past. The use of differential taxation is currently used to stimulate greater fuel efficiency but, following the lead of the EU[18], the Government is looking at ways of extending this. There are also measures to reduce the amount of fuel supplied to employees (p.122).

What do we know from Hindsight?

Looking back is not only interesting it is also important for learning lessons. The 1968 Transport Act was highly influential and many parts of it were enduring and very successful. The removal of quantity licensing from road haulage is the classic example of a success as was the creation of, the subsequently privatised, National Freight Company[19]. It also helped to reduce the reliance on indirect, cross-subsidies as a tool for ensuring the provision of social public transport services. The 1960s also laid the basis for a significant research presence in transport in the UK and resulted in better official data collection methods and improved statistical analysis.

Where it was less successful was in coping with the long term problems of the decline in the railways and other forms of public transport (save aviation). The notion of specific rather than block subsidies that prevailed was innovative but proved practically very difficult. The measures of integrated urban transport and land-use policies did little to cope with increasing traffic congestion. Indeed, the efforts to encourage people to use public transport were singularly disappointing. The lure of the motorcar is much stronger than most transport models predict. Many efforts to foster public transport use ignore the massive quality differences involved in getting straight into a car at one's home after being able to choose exactly when to leave the house, adjusting the temperature to that desired, listening to music that one likes, enjoying one's own company or company of one's choosing, being able to conveniently carry large packages in security, to select alternative routes periodically to avoid boredom, sitting in seats that have been selected from the wide range that are available in different vehicles and so on. Public transport does offer an attractive variation on dry days when the timing is convenient and one has little to carry but in most cases there are major qualitative differences in service quality.

The new White Paper recognises part of the difficulty in moving people from their cars (pp.40-41) but the experiences of the past indicate a significant stick is required to get people out of their cars. The White Paper hints at deploying some sticks, but the lesson still does not seem to be fully learned. Improved public transport may be nice for those who currently use it and may help to retain their patronage, but in many smaller towns and cities, public transport is still likely to remain unattractive to car owners. While it is nice to make firm commitments about carrots, sticks are less attractive and can be put off for another day. This seems to have been the standard approach to transport in the past and remains so in this document.

Governments almost seem to believe their *raison d'être* is to set up official bodies. In the 1960s, for example, there were the Passenger Transport Executives and Authorities. Now we have the Commission for Integrated Transport. Official bodies seem to have one of two functions. They either try to manage something (the 1960s model) or to offer advice (the 1990s model). The 1960s model was not an outstanding success. There are, however, actually some very useful advisory bodies; SACTRA is an example and, more generally, so is the Royal Commission on Environmental Pollution. One feature that seems important in determining success is that the body should be focused. The main problem with the notion of a Commission for Integrated Transport, judging by the topics it is meant to be advising on (pp.92-3), is that it has no focus. It is meant to acquire expertise and dispense advice on traffic forecasting (some skills in discrete choice modelling presumably being a prerequisite for membership), on rail freight (a Ph.D. in logistics clearly being essential), to get the best value from public subsidies (chartered accountancy being useful here) and on cost-benefit analysis (this involves economics so no real training is necessary). It is interesting to know what this will add to the role of the relevant House of Commons Select Committee or the internal expertise of the Department.

The Strategic Rail Authority appears to be more focused and with clearer objectives but there are vexing institutional issues of actually developing an agency, 'combining pragmatism with a strategic view', (p.97). How it is to be structured and its detailed terms of reference are not addressed in the White Paper but striking a balance between surrendering to the needs of short-term crisis management concerns is not unknown for a supposedly strategic institution. Focusing on longer term issues will not be easy.

Conclusions

The recent transport White Paper represents another stepping stone in the way transport policy is evolving in the UK. It is comprehensive in its coverage and follows a pattern established in the 1960s of treating transport holistically. As a document it is by-and-large remarkably well written and presented and reflects the continuing concern with Governments of all political persuasions to communicate with their constituents (even the infants in this case). It also picks up the national and international trend to embrace environmental considerations more completely into sectorial policy formulation; it also does so in a relatively balanced way.

The document is much less important in terms of the policy details that it contains than in terms of the clarity with which it paints the nature of the current problems confronting policy makers and of the larger geographical arena in which UK transport policy must be developed. In this context it contains nothing that is really revolutionary but the arguments are well expressed.

An interesting question is whether the current package of proposals will retain an enduring position in the annals of transport history in the same way the Labour Party's policies of the 1960s have done. One suspects not. The documents itself flits too frequently from the big field to the niche interest in its apparent bid to please as many groups as possible. In that sense it lacks the authority of *Transport Policy*. The theme of integration is also too general and vague to be enduring and the policy measures, while perhaps useful, are not genuinely innovation.

References

British Railways Board (1963) *The Reshaping of British Railway* HMSO, London.

Button, K.J. (1974) Transport Policy in the United Kingdom: 1968-1974, *Three Banks Review,* 103, pp.26-48.

Button, K.J. (1982) *Transport Economics* 1st ed. Heinemann, London.

Button, K.J. (1992) The Liberalisation of Transport Services In D.Swann (ed.) *1992 and Beyond* Routledge, London.

Button, K.J. and Gillingwater, D. (1986) *Future Transport Policy* Croom Helm, London.

Button, K.J. and Keeler, T. (1993) The Regulation of Transport Markets *Economic Journal,* 103, pp.1017-28.

Button, K.J. and Swan, D. eds. (1989) *The Age of Regulatory Reform* Oxford University Press, Oxford.

Commission of the European Communities (1992) *The Future Development of the Common Transport Policy: A Global Approach to the Construction of a Community Framework for Sustainable Mobility* CEC, Brussels.

Commission of the European Communities (1995) *Towards Fair and Efficient Pricing in Transport* CEC, Brussels.

Gerondeau, G. (1997) *Transport in Europe* Norwood Artech House.

Munby, D.L. (1968) Mrs Castle's Transport Policy *Journal of Transport Economics and Policy,* 2, pp.135-73.

Organisation for Economic Cupertino and Development (1994) *Competition Policy in OECD Countries* OECD, Paris.

Socialist Commentary (1975) *Transport Policy: The Report of a Study Group* Socialist Commentary, London.

Starkie, D. (1984) Policy Changes, Configurations and Catastrophes *Policy and Politics,* 12, pp.71-84.

UK Department of the Environment, Transport and The Regions (1997) *Developing and Integrated Transport Policy. An Invitation to Contribute* HMSO, London.

UK Department of Environment, Transport and The Regions (1998) *A New Deal for Transport: Better for Everyone. The Government's White Paper on the Future for Transport* Cmnd. 3950, The Stationery Office, London.

UK Ministry of Transport (1964) *Road Pricing: the Economic and Technical Possibilities* HMSO, London.

UK Ministry of Transport (1966) *Transport Policy* Cmnd 3057, HMSO, London.

UK Ministry of Transport (1967a) *Railway Policy* Cmnd. 3459, HMSO, London.

UK Ministry of Transport (1967b) *The Transport of Freight* Cmnd. 3470, HMSO, London.

UK Ministry of Transport (1967c) *Public Transport and Traffic* Cmnd. 3686, HMSO, London.

World Commission on Environment and Development (1987) *Our Common Future* Oxford University Press, Oxford.

Notes

1 Page numbers in parenthesis with no other point of reference refer to material in UK Department of Environment and Transport, (1968).

2 Indeed, the 1968 Transport Act's successful deregulation of road haulage preceded many reforms of transport regulation elsewhere in the World, such as the 1978 Airline Deregulation Act in the US, which are often held up as the key turning points in regulatory liberalisation.

3 For example, the British Airports Authority is the subject of price capping.

4 While this position receives a lot of attention, it is not universally held, e.g. for a counter view see Gerondeau, 1997.

5 The same broad message is found in the Conservative Government's policy document, *Transport: the Way Forward* and is acknowledge in the new policy document (p.12). In practice there was rhetoric concerning urban traffic restraint in the 1970s, largely due to local objections to the expansion of urban motorway systems, but the scale of the policy response was comparatively limited.

6 E.g. the benefits seen from the 1978 Airline Deregulation Act provided powerful demonstration effects for liberalisation elsewhere in the world.

7 As pointed out in Roosevelt's inaugural speech in 1932.

8 E.g. '...all the transport matters for which local authorities are to be responsible...must be in an integrated transport plan' (UK Ministry of Transport, 1967c, para.10), also 'full integration of the planning of road and rail services' was seen as a priority (UK Ministry of Transport, 1967a).

9 The underlying philosophy of the Labour Party's transport policy in the mid-1970s is outlined in Socialist Commentary (1975). Button (1974) offers a discussion of the policy between 1968 and the return of the Labour Party to power in 1974.

10 The need to conform to EU maximum axle weight laws for lorries being a clear example (p.71).

11 Although there is the inevitable note of caution that always seems to accompany any idea of making greater use of the charging mechanism as a policy tool - e.g. 'In designing further [charging] projects we will consider what lessons can be drawn from overseas' (p.116); 'Further studies are required on the electronic units and on administrative support systems' (p.116), etc.

12 The omission possibly reflects the age of the policy makers and the absence of a co-ordinated roller-blading lobby.

13 The data given also relates to average figures rather than those for the days of worst weather in each case.

14 These included such studies as the Buchanan (1963) report on urban transport., the Smeed Report on road pricing (UK Ministry of Transport, 1964) and the Beeching Report on the railways (British Railways Board, 1963). It is worth noting that, not only did studies appear more personal in the past, giving a target to aim at, but many of these studies were legacies of the previous Conservative government as are some of those cited in *A New Deal for Transport.*

15 Perhaps the most important of the policy assessments were UK Ministry of Transport (1966; 1967a,b,c).

16 The main consultation documents were UK Department of the Environment, Transport and the Regions (1997a,b).

17 E.g., regarding the car, 'It has brought...noise, fumes and danger as a setting of our lives; a rising trend of casualties on our roads and a threat to our environment in both town and country' (UK Ministry of Transport 1966 para 1).

18 E.g. as outlined in Commission of the European Communities (1995).

19 The NFC only really revealed its value after privatisation in 1982.

3 Charging for Access to Railtrack's Network: Some Current Issues

CHRIS BOLT

Introduction

In the recent White Paper *Fair Payment for Infrastructure Use: A Phased Approach to a Common Transport Infrastructure Charging Framework in EU*, the European Commission (1998) set out its proposals for developments in infrastructure charging, designed to improve the overall efficiency of the provision and use of European transport infrastructure, promote competition, safeguard the single market and enhance the sustainability of the transport system. It identified four basic principles on which such a common charging framework should be based.

The EC White Paper went on to identify a number of specific changes needed to deliver these principles in different modes. In the case of railways, it has also published draft directives to implement these proposals.

This paper reviews the EC principles, and their relevance to the current review of the structure and level of Railtrack's access charges. In the light of that assessment, it considers how far the current framework of Railtrack charges, established in 1995, meets those principles, and also assesses the detailed proposals put forward by the Commission for changes in rail infrastructure charges. There are potentially significant issues given that EC Directives will, once implemented, override the provisions of existing domestic legislation. The paper suggests an alternative approach, which, in the author's view, more appropriately reflects the public/private partnership that is increasingly characterising transport infrastructure networks.

The EC Principles

1.   Consistency Across Modes and Across Member States

The EC's four principles were:

(a)   The same fundamental principles should be applied to all commercial modes of transport in each Member State of the European Union, while recognising that the resulting structures may differ by mode and the level of charge may differ by location to reflect different needs and circumstances.

(b)   Infrastructure charges should encourage greater efficiency in the use of transport infrastructure and, therefore, be based on the 'user pays' principle: all users of transport facilities should be charged for costs they impose at, or as close as possible to the point of use.

(c)   Charges should be directly related to the costs that users impose on the infrastructure and on others, including the environmental and other external impacts caused by the users.   Charges should only differ when there are real differences in costs and service quality and should not discriminate between users on the basis of nationality and residence/business location.

(d)   Charges should promote the efficient provision of infrastructure.

In a simple text book world, it is straightforward to demonstrate that a marginal costing approach is appropriate to ensure economic efficiency, provided any economies (or diseconomies) of scale can be dealt with through a framework of optimal taxation.   But the real world does not match the textbook.   The critical question this then raises is how far the 'second best' framework for infrastructure charging departs from a marginal costing paradigm.   In particular, given that charging frameworks are in practice consistent neither between modes nor between member states, is one type of consistency more important than another?

• Underpinning this question is the caricature that rail transport is 'better' than road transport, for both passengers and freight, but requires subsidy because charges for road use are inadequate. There is no doubt that this is a caricature;

• as David Newbery argues in Chapter 3, road users in the UK more than cover their direct costs through vehicle and fuel taxes, and probably cover social costs as well;

• although road pricing has yet to develop, fuel taxation means that road use is not strictly speaking 'free at the point of use' (although the pricing signals may be unclear as well as imperfect); and

• even in environmental terms, rail is not unambiguously better than road; much depends on loadings.

So in the context of 'competition' between modes, it is far from obvious that the current charging framework necessarily disadvantages rail. This is perhaps even more true now that the assessment of changes in rail services in the UK is clearly placed on a cost benefit framework (OPRAF, 1997). It also suggests that claims that the decision to abolish access charges for the

Swedish rail network place it on a 'level playing field' with road are far from valid.

The issue of geographical consistency also needs to be treated carefully. This is particularly true for the United Kingdom with only a single rail link to other member states. But even for member states sharing land boundaries, constraints on growth in rail traffic and competition between different operators may have much more to do with questions of access and timetabling - and also technical standards - than with inconsistencies in charging systems. The current 'freight freeways' initiative seeks to overcome these problems by establishing a 'one stop shop' for freight paths on specified routes.

2.   *The 'User Pays' Principle*

Just as with consistency, the principle that users should pay for the costs they impose, in order to encourage greater efficiency and use of infrastructure is on the face of it sensible, but should not be pushed to extremes. The practicalities of charging systems need to be taken into account to avoid a situation where the costs of the charging system itself dominate the total charge. For example, it is part of the current charging framework that Railtrack's costs of electricity current for traction should be recovered only from operators of electric trains. To go further and relate these costs directly to the electricity consumption of individual trains, would require meters in cabs. The arguments here are a bit like water metering for households; costs are much lower if meters are installed at the time of construction, and the case for retrofitting depends crucially on the potential for avoiding more expensive new water resource developments. It is not obvious that the costs of retrofitting existing locomotives with electric meters would be cost effective in terms of the resulting cost savings.

3.   *Costs Should Include Environmental and Other External Impacts, With Charges Established on a Non-Discriminatory Basis*

This principle contains two elements. The notion that user costs should be established on a basis which reflects environmental and other external impacts is in essence related to proposition 2. and the same argument about measurement and cost effectiveness of charging systems apply.

Discrimination is a more problematic issue. The approach taken in the EC White Paper is that any difference in charges which does not reflect differences in supply (i.e. differences in costs or in the product being supplied) amount to discrimination.   There is, of course, extensive economic literature on the merits of charging systems - such as Ramsey pricing - which depart from this approach in order to increase economic efficiency in the context of a budget constraint (i.e. the non-availability of optimal general taxation).   In the context of Article 86 - the prohibition on abuse of dominant positions which has been incorporated into UK law through the 1998 Competition Act provisions - case law supports the conclusion that differentiation in pricing between customers or categories of customers on the basis of matters in addition to cost is not precluded. The key issue here is whether additional factors can be objectively determined and whether differentiation in prices to customers will have an impact on competition in final markets.

4.   Promoting the Efficient Provision of Infrastructure

It is, perhaps, worth identifying a number of aspects of efficiency.   One is 'productive' efficiency - incentives to deliver infrastructure at a cost on the efficiency frontier.   Another is 'allocative' efficiency: delivering the right type and amount of infrastructure.   Again, in practice, it may not be easy for practical charging systems to focus on both, and a decision will be needed on which objective is more important.

Overall, it is difficult to argue too much with the principles as stated. But they are not of themselves particularly helpful in terms of developing practicable charging systems.   As discussed later, they can lead to proposals which do not fit well either with the UK railway structure or with sound economic principles.

The Current Structure of Railtrack's Access Charges

The current level and structure of Railtrack's access charges were put in place from 1 April 1995 following a review by the Rail Regulator in the second half of 1994.   The initial structure of charges reflected decisions taken by Government on the basis of its policy statement *Gaining Access to the Railway Network*, February 1993.   The main features were as follows:
- the overall level of charges was set at a level to allow Railtrack a reasonable return on its assets (the key difference between the April 1994 charges and the April 1995 charges being whether this should be a replacement cost asset value, or some lower value);

- charges were set on a 'single till' basis, in particular with property income used to offset access charges;
- within the overall level of charges, freight services were expected to cover total 'freight specific costs' (i.e. those costs which would be avoided if the network was passenger only);
- all costs common to passenger and freight were effectively recovered from passenger services;
- the initial endowment of access rights for passenger services was charged variable charges to reflect (short run) wear and tear track usage costs and traction current costs, with remaining costs, determined by the Regulator in accordance with the duties of section 4 of the Railways Act not to make it unduly difficult for Railtrack to finance its activities, recovered as an annual lump sum from individual train operators, on the basis of various attributions and allocations;
- negotiation of charges for individual freight access agreements and for additional passenger access rights, over and above the initial endowment; and
- total subsidy levels were determined essentially as an output of the initial franchising process, rather than being an input.

In addition, where passenger operators have sought additional access rights over and above the initial endowment, these have been negotiated on a basis which covers at least Railtrack's avoidable costs, with a share of any benefits depending in part on the risk allocation between the parties. In some cases, such as the West Coast Main Line upgrade, the commercial terms have included a formal revenue sharing arrangement.

The current charging structure therefore follows different principles for different elements of Railtrack's business:

- initial passenger services are essentially on a short run avoidable cost basis, with an 'administered' annual fixed charge for each train operator;
- freight services are also on a similar basis, but with the fixed charge designed to ensure recovery of total freight specific costs; and
- additional passenger services are expected to pay charges which in principal cover long run avoidable cost, with the possible addition of a share of net benefits.

The Regulator recognised, in reaching decisions on the structure of access charges, that his proposals were far from ideal. But both time and the availability of relevant costing information precluded more significant changes to the structure put in place by Government in April 1994. With hindsight, however, it is easy to see how far the current structure departs from appropriate economic principles - and certainly the principles advanced by the EC.   Even in matters of terminology, for example, consistency has been lacking.

When the Regulator reached conclusions on the current structure of charges in November 1994, he recognised both the constraints on changes that could be made in the short term, but also the need to put in place arrangements for a more fundamental review in due course. In particular, the Regulator endorsed the proposition that, with on average over 90% of the access charge in the form of fixed sum, it was desirable to achieve greater variability.

These concerns were restated when the Regulator launched the periodic review of Railtrack's access charges in December 1994. Indeed, one of the reasons given for the early start to the review was the long lead time needed to address structure of charges issues.

The first periodic review consultation document (ORR, 1997) put the issue in the following terms:

> Over time the Regulator has become concerned that these advantages have been outweighed by the disadvantages. In particular he has become concerned that, while the existing structure might encourage the use of the existing network, it has not been successful in encouraging Railtrack to make incremental enhancements to the network. He is concerned that the transaction costs associated with these relatively frequent but financially small negotiations are high. He has also become concerned that information asymmetries and in particular the lack of understanding on the part of train operators about the costs which Railtrack incurs may in total be leading to Railtrack receiving an excessive return.   On a number of occasions the Regulator has been required to reject modifications to access agreements on the grounds that Railtrack is sharing in an excessive amount of the benefit which the incremental service generates, in view of the risk which it takes in providing the access.
>
> The Regulator is also aware that the variable charge which is currently included in access agreements for track usage does not reflect the cost of increasing capacity on those parts of the network which are congested. The effect of this is to remove economic signals about where demand for developing the network is sufficient to justify expansion, and would guide efficient use of the existing capacity where it is scarce. Increasingly, the Regulator has found that he has been required to take administrative decisions about the allocation of train paths on parts of the network which are being operated close to or at capacity.

The Regulator subsequently confirmed his intention to take forward this element of the periodic review programme. As part of this process, the Regulator and Railtrack held a seminar in April 1998, on which the proceedings have subsequently been published [reference http://www.rail-reg.gov.uk], and have jointly drafted a paper setting out the principles which a revised structure of access charges might be expected to reflect. The proposed regulatory objectives and characteristics of an appropriate charging framework are at Annex A.

The EC 1998 Proposals for Rail

The EC White Paper develops proposals for common charging systems on the basis of its four principles and a 'practical definition of marginal social costs' as follows:

> Marginal costs are those variable costs that reflect the cost of an additional vehicle or transport unit using the infrastructure. Strictly speaking, they can vary every minute, with different transport users, at different times, in different conditions and in different places. Moreover for the last extra carriage on the train, car on the road, or ship at sea, marginal costs can often be close to zero. Clearly such a strict degree of approximation and averaging is necessary to develop understandable, practical charging structures. Marginal costs may at times merely reflect an average of variable costs. More usefully, they should reflect infrastructure damage, congestion and pollution costs, and so would vary according to factors like unit weight or number of axles, peak times, urban travel, and engine emissions.
>
> Marginal cost components can include:
> - Operating costs: energy, labour, some maintenance costs. Infrastructure damage costs: maintenance costs, wear and tear of the infrastructure, reflected by such as resurfacing of roads, rails and runways.
> - Congestion and scarcity costs: The cost of time delays to other users or non users, resulting from congested traffic flows (on roads, queues for airports or railway stations). Moreover, a transport operator's use of infrastructure may prevent another operator from using it (e.g. an airport runway).
> - Environmental costs: air, water, and noise pollution.
> - Accident costs: Costs in terms of material damage, pain and suffering and production losses.

The specific rail proposals are contained in section II of the draft proposed directive which it is envisaged might replace the current directive 95/19/EC. The main features are:

- a general presumption of short run social marginal cost pricing (i.e. including congestion and environmental costs but not capital costs (Article 8);
- inclusion of 'long run' elements in narrowly defined circumstances, and in any case only for passenger services (Article 9);
- discounts allowed where the costs for a package of services is less than the sum of the charges for individual components (Article 10);
- encouragement of a performance regime (Article 12); and
- the possibility of reservation charges (Article 13).

Although this framework has some features in common with the existing Railtrack structure, for example the short term nature of variable charges, it clearly differs in a number of important respects.

The issues can, perhaps, be best illustrated with an example. Consider a freight-only part of Railtrack's network, used by two freight services. The costs avoidable in the long term if the route is closed is 100 (in present value terms); the costs avoided if either of the two services ceases is 10. It is also assumed that there are no prospects of additional freight (or passenger) use.

The EC rules suggest that the appropriate (and the maximum) charges to both users are 10 (assuming the line is uncongested, and that there are no environmental costs). This also would appear to be the implication of the current Railtrack charging system, which argues that any flow which can at least meet its avoidable costs should be allowed on to the network.

The critical question, however, is how Railtrack recovers the residual costs of 80, if each user is only charged 10. In particular, how is this achieved in a way which does not involve cross-subsidy between passenger and freight, or undue discrimination between different freight users?

In answering this question, it is important to recognise that, as a privately owned company, Railtrack will, quite properly, react on the basis of the commercial incentives it faces. The objective of regulation (and Government subsidy) is to ensure that those commercial incentives properly reflect the underlying public interest.

Let us start by considering the position where the two freight flows are essentially identical (i.e. there are no objective demand factors distinguishing them). In that case, each freight flow should pay an access charge of 50; if this charge cannot be afforded (even taking into account any track access grant from Government, reflecting environmental benefits of rail freight), then Railtrack should close this part of its network. (In practice, there may be timing issues here. If costs are lumpy, it may be economic to keep operating the line until major renewal costs are incurred.) If the charges were different to the two users, then competition in final markets might be distorted.

Where, however, the two flows have different characteristics - say one is coal, where rail freight has comparative advantage, and the other is a more marginal mixed traffic flow - then different charges may be appropriate. But even here, it might seem reasonable to argue that the two users should still be expected to pay between them a charge of at least 100; if they did not, avoidable costs attributed to one part of the network would be borne by users on another part. Those other users would clearly be better off if the uneconomic part of the network was closed.

This example can be extended in a number of ways. For example, if the two users are in the same market, but the quality of access rights differs, charges would again be different. This is, however, permitted at least to a degree under the EC proposals.

What does this mean in practice? A requirement to charge freight operators short run marginal social cost, with no attempt to assess the longer term costs and benefits of rail freight versus other modes clearly risks distortion of competition between modes. But in the case of Railtrack, a privatised company entitled to - and needing to - earn a reasonable return on its assets, the distortions are even greater. It is not, for example, simply a matter of subsidy being paid to recover the shortfall in total freight costs; if the draft directive were implemented it would require the Regulator to endorse cross-subsidy between passenger and freight services. It would also quite possibly require the Government to pay subsidy direct to Railtrack, rather than supporting particular services through subsidy payments to train operators. Nowhere does the EC proposal justify this policy outcome.

In respect of the existing network, where the regulatory asset value is considerably below replacement cost, this may not in practice be a major distortion. But where network enhancement involves provision for freight, a prohibition on recovering more than the short run avoidable cost would distort - and reduce - investment incentives, and risk frustrating the very objective of increasing capability for rail freight.

The draft EC proposals therefore contain a fundamental contradiction. The ECs approach to transport, and rail in particular, is based on the need to promote a public sector/private sector partnership; private sector operating skills and private capital, but where socially desirable services are not viable, public money to support their provision. Such a public sector/private sector partnership characterises the new structure of the railways in Great Britain.

While overall EC policy is in support of such partnerships - the 1996 EC White Paper called for a 'new kind of railway' - its detailed policy description for charging remains very much in the mould of a public sector industry. No account is taken of the need for an infrastructure company to earn a reasonable return, either on its existing assets or, where freight is involved, on new investment. Nor is proper account taken of the need for a consistent 'second best' across modes.

An Alternative Way Forward

As part of its work on Railtrack's periodic review, ORR has agreed with Railtrack a detailed work programme which starts with Railtrack assembling appropriate information on cost causation, both in respect of the existing network and in terms of enhancement schemes. Such information is essential if changes in the structure of charges are to be robust.

The output of railway infrastructure is complex. It is not simply a matter of 'a train path': a path into London Victoria in the commuter peak is clearly very different from an off-peak path on the heart of Wales line - or even, for that matter, an off peak path into Victoria. So various dimensions of access can be identified, including: journey times; reliability; firmness of rights (e.g. 'clockface' departures); time of day; and geography.

This heterogeneity highlights a fundamental dilemma: can a sufficiently small number of characteristics be identified as a basis for a 'tariff' for access charges? Or are the differences so wide that any such attempt will lead to within-group variances bigger than between-group variances? If the practical research on cost causation suggests a reasonably small number of parameters can be identified to explain the major variations in costs, and that a tariff approach is feasible, then there is a prospect of significant development in the current approach to access charging, with its mix of administered fixed charges for the initial endowment of access rights and negotiation for additional rights.

For example, Railtrack's network might be divided up into the following broad categories, each with a different approach to charging for bundles of access rights:

- those where, for the foreseeable future, the long run incremental costs of changing asset capability are likely to remain above users willingness to pay for the changes concerned;

- those where changing the capability of the assets is best organised as a major, co-ordinated programme of works to meet a complex and interdependent array of user requirements (such as the upgrade of the West Coast Main Line); and
- those capable of smaller scale enhancements and alterations, the costs of which are lower than or equal to users' willingness to pay for them.

For the first category, charges reflecting short run avoidable costs would be appropriate. These would include an element to reflect the impact on Railtrack's costs through performance regimes of worsened overall train performance on congested parts of the network.

The second category is more suited to negotiated charges which ensure that risks and incentives are properly allocated to the parties best able to manage them; in this regard, standardised tariffs have less of a role to play in signalling investment but may still be useful in managing the subsequent operation of the assets.

For the remaining assets, network changes will typically occur in smaller lumps, and enhancement investments are typically combined with renewals spending triggered by asset condition and general standards (rather than specific user demands) around the network. In these circumstances, charges which identify the relevant long run incremental costs of changing network capability can provide sustained and steady signals to infrastructure managers and users for the efficient ongoing development of the network. Again, there is an analogy here with the water industry, and the 'Paying for Growth' principles developed by Ofwat.

Such a mix of short run and long run charging would be aimed at giving improved incentives for Railtrack to manage the use of scarce capacity, while providing the financial resources for network enhancement where operators were willing to pay the appropriate long run incremental cost. It would also allow an approach where a future Strategic Rail Authority met some or all of the capital costs of enhancements which were socially desirable, but not commercially viable, but not necessarily down to the point where users paid only short run marginal cost if that would give unreasonable profit to users.

If a matrix of costs, and hence charges, can be developed on this sort of basis, two important issues remain. One is whether charges should simply be levied on trains operated, or whether some should be on the basis of contracted capacity. The other is how the remainder of the regulated revenue requirement should be recovered. But if variable charges are higher than the current level (less than 10% of total charges) and the total revenue requirement is tightened (for example through the use of a regulatory asset base related to the market value of Railtrack at the end of the first day's trading), then the scale of the difference will be very much less than in 1995.

Proposals for a revised structure of charges will be published by the Regulator in Spring 1999.

References

European Commission (1998) *Proposals for Directives Concerning Infra-structure* COM(98)480 22/7/98 European Commission, Brussels.

OPRAF (1997) *Assessment of Passenger Rail Services: Planning Criteria: an Interim Guide* OPRAF. London

ORR (1997) *The Periodic Review of Railtrack's Access Charges - A Proposed Framework and Key Issues - A Consultation Document*. ORR, London.

Annex A

Railtrack's Access Charges: Economic Pricing Principles, Implementation Principles and Issues

Extract from ORR/Railtrack paper

Regulatory Objectives

In considering the structure of track access charges, the Regulator's main objectives are to develop a framework of charges which:
- incentivises Railtrack, train operators and funders (OPRAF, DETR, and PTEs) to maximise the efficient use and development of the network;
- does not unduly discriminate between users of the network;
- delivers the appropriate overall level of charges consistent with the periodic review process and does not over or under value the reward to Railtrack for changes in the level of output[1];
- best meets the Government's overall transport objectives, such as promoting the use of the railway network in Great Britain; and is practical and effective in its operation.

Characteristics of an Appropriate Charging Framework

The framework of track access charges that best meets the above objectives should have the following key characteristics:
- comprehensibility - the structure should be understood by the industry participants whose behaviour it is meant to influence and should not impose undue transaction costs to identify the appropriate information;
- transparency - the structure should provide clear information to industry participants on the make-up of charges, and hence not confer undue advantage on particular industry participants through information asymmetry;
- stability - charges should not fluctuate or alter in arbitrary or unpredictable ways, except where significant short term cost changes are being signalled - if congestion (scarcity) pricing is introduced, short run prices could be unstable but predictability about future average

levels could be given in some cases by establishing a long run avoidable cost around which short run prices might be expected to fluctuate;

- measurability, cost effectiveness and objectivity - the data required to derive charges should be objectively measurable, cost-effective to collect and unambiguous to apply (for billing purposes); and

- cost reflectivity - in order to meet the Regulator' efficiency and funding objectives for charging structures, charges will need to be cost-reflective. The economic interpretation of these requirements, in the context of other practical requirements for Railtrack's track access charges, is discussed in more detail in the remainder of this paper.

Note

1   Output here and elsewhere refers to whatever will change under the decision at issue, for example, numbers of trains, timing of trains, train speeds, axle loadings, use of regenerative braking, withdrawal of service, etc.

4 Road Pricing and Road Finance

DAVID M NEWBERY[1]

Introduction

The title of the recent Transport White Paper *A New Deal for Transport: Better Transport for Everyone* captures the flavour of a document whose message is that everyone will be better off in the best of all possible transport worlds. Efficiency in road transport requires both that the level of service is adequate and that scarce road space is allocated efficiently. There is a welcome recognition that economic instruments including road pricing might help the second of these objectives, but there is little discussion of the problem of providing and financing an adequate level of provision, and what little there is suggests fudge rather than rational policy making.

Table 1   Road costs and taxes for 1996/97

	£ billion
Revenues from road taxes	
Fuel Tax	17.2
Vehicle excise duty	4.2
Total tax revenue	*21.4*
Cost of road provision	
Interest on capital at 6%	7.2
(Capital expenditure)	*(3.3)*
Maintenance, policing, etc.	3.9
Total road costs	*11.2*
Surplus of revenue over cost	*10.2*
PCU km (billion)	483 billion km
Cost per PCUkm pence/km	*(2.1 p/km)*
Road taxes per PCUkm pence/km	*(4.4 p/km)*

41

Table 1 shows that motorists contributed over £21 billion in road taxes in 1996/97, almost double a generous estimate of what it cost to provide the services they consumed, including 6% real return on the capital value.

The 1998 budget has increased various fuel taxes by between 9.2% and 11.8% compared to 1997, so that tax revenue will increase by 16% over the two years from the figures in Table 1. The excess of tax revenue over charges might be £14 billion or 2.8p per kilometre. The *increase* in road tax revenue over these two years is thus comparable to the *total* amount earmarked in the *White Paper* for funding road transport.

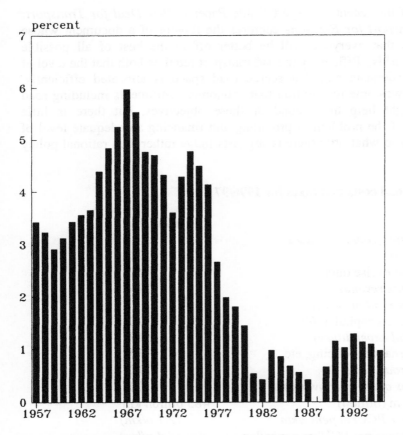

Figure 1 Net public sector capital formation as share of GDP
Source: Blue Book 1996 (Psinvest)

The *White Paper* lays out its views on the financing of transport in its chapter 4 (parts of which are reproduced in the Appendix). This starts promisingly with a commitment to doubling the level of net public investment as a share of GDP. This sounds dramatic but so low has net public investment fallen that this would only increase it from about 1% to 2%, as Figure 1 shows, and would only involve returning gross levels of investment to their average from 1988-95, as Figure 2 shows.

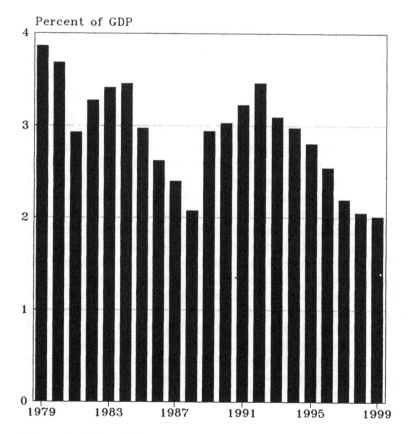

Figure 2 UK public sector capital spending excluding industries now privatised
Source: Treasury PSBR 1997

The chapter also states that the level of privately-financed investment in transport will increase by at least a half over the next three years, which is worrying, given the criticisms levelled at this expensive form of finance by the Audit Commission, and again suggests that the Government is seeking to avoid its responsibility for ensuring adequate investment. On the other hand there is the encouraging statement that the public spending rules have changed, and that 'the annual spending round which encouraged short-termism and inefficiency' (§4.11) has been abolished. Whether the proposed spending limits for the next three years will in fact give greater certainty and stability remains to be seen, particularly what happens in the third year, and when all the commitments to finance the Private Finance Initiative schemes (a large fraction of which are in the transport sector) come home to roost.

The disappointing aspects of transport and especially road finance have to do with the methods of finance that are ruled out, those that are not mentioned, and the ambiguity surrounding those that are included. Thus chapter 4 states that 'Although transport investment at national level could be funded by dedicating particular streams of taxation income, as some suggest, that would inevitably restrict our ability to use that income flexibly both for transport and for other priorities such as education and health.' (§4.9). This of course raises the immediate worries that the promised stability in infrastructure investment will be put at risk by the need to switch funds to nurses, or new drugs, or reducing hospital waiting lists, and casts doubt on the value of the three year commitment.

The other implication of this statement is that this Government has ruled out a radical reform of road funding of the kind set out most recently in Newbery (1998b). That lecture argued that two recent developments had potentially transformed the management and finance of public sector infrastructure. First, experience in regulating privatised network public utilities like gas, water, and recently Railtrack, had demonstrated the feasibility of designating revenue streams for long-term programmes of asset renewal and management. Secondly, the Government earlier announced its commitment to the Golden Rule, which states that 'over the economic cycle, the Government will borrow only to invest and not to fund current spending'. This can be rephrased to state that investment can be financed by borrowing, and it is not necessary to resort to PFI for infrastructure investment. Taken together, this makes it feasible and desirable to fund road transport out of part of the current road taxes, which would become road user charges, under the regulation of an agency that might be called the Office of Road Regulation or Ofroad.

Not only is such revenue designation ruled out, but there is no mention of the Golden Rule as a reason for relaxing current account spending limits, perhaps because of a reliance on PFI (which has the same macro-economic impact) rather than on public sector borrowing. Instead, we are offered road pricing as a carrot to encourage local authorities to finance transport improvements: 'We will therefore introduce legislation to allow local authorities to charge road users...' (§4.94). 'Schemes may be developed, for example, to help to meet transport and environmental objectives in urban or rural areas..... We will also consider for each scheme how best revenues generated may be used to provide related benefits locally - which might otherwise be unaffordable, including better means of securing the environmental acceptability of transport infrastructure.' (§4.100).

Part of the motive for introducing such road pricing is the laudable aim of improving the efficiency of road use, which is discussed below, but part is clearly the aim of providing an additional revenue stream for financing improvements in local transport systems. There is no discussion of the problem of dual funding, where some money is voted by the Central Government, and the balance is raised from road pricing, and there is no clear discussion of the need to regulate the setting of road prices. The *White Paper* announces the creation 'of a new independent body - the Commission for Integrated Transport (CfIT) - to provide independent advice to Government on the implementation of integrated transport policy, to monitor developments across transport ..' (§4.4). This new body appears to be solely advisor, with no regulatory functions. It is given an impressive list of advisory tasks, but they do not include monitoring the charges levied by local authorities which will be available for local transport investment.

In the past, several Mediterranean countries used to operate a system of Octroi whereby travellers would be charged a tariff on their goods every time they entered a town. Needless to say this was a major impediment to economic development and trade, and raises suspicions that local authorities might be tempted to raise tolls on out-of-town motorists to finance benefits for the local community unless they were subject to close regulatory oversight or restraint.

There is a second worry about the new revenue raising powers that are to be allocated to local authorities. To the extent that they were successful in financing local transport investments, the Central Government might then reduce its allocation of funds, arguing that transport needs were being satisfactorily met. Unless there is some method of defining the amount of centrally funded expenditure (which is surely best done by designating revenue streams) or of demonstrating that local road pricing is truly additional revenue, all confidence in the accountability and stability of transport funding will be lost.

Road Pricing to Improve Efficiency

The concept of treating road transport as a public service like rail transport suggests that it should be priced both to raise the required level of revenue (to be determined by Ofroad once the Government has decided on the level of transfers that reflect the asset value of the road system) and to achieve efficiency. Most UK public utilities had modern equivalent asset values that greatly exceeded their market value when privatised, and the regulators then had to determine the regulatory asset base and the allowed rate of return, which, with the estimated operating costs, determines the level of the price cap sufficient to finance operations and reward owners[2]. The solution adopted was to write off the difference between the market value and the correct asset value and treat it as effectively a windfall to consumers (in the case of water and gas), or a subsidy towards meeting social obligations (in the case of Railtrack). Fortunately, in the case of roads, there is little problem in setting a return to the Treasury (as owner) on the estimated value of the network without writing it down, so large are current road taxes. Thus the problem of pricing when existing capital has a low average cost but new investment has a higher marginal cost would not arise for road transport.

If there is no problem in generating sufficient revenue from the current level of road taxes, it is clear that the present system of road taxes does not allocate scarce road space very efficiently. The attraction of road pricing is that it offers the prospect of improving on the present rather blunt instruments of fuel taxes and VED. Table 1 showed that fuel taxes were 4.4p/PCUkm in 1996/7 and will now be over 5p/PCUkm. The marginal social cost of congestion in urban areas might be seven times as high as this in the rush hour, but on most inter-urban and rural roads it will be a small fraction of this level. Even if fuel taxes double over the next decade (as they are set to do at present) vehicles will be undercharged on busy urban streets, but would be even more seriously overcharged elsewhere.

The *White Paper* cites the Royal Commission on Environmental Pollution's *Report on Transport and the Environment: Developments since 1994*. There the claim is made that 'forecast traffic growth is economically, environmentally and socially unacceptable'. The *Report* recommends making maximum use of capacity of existing trunk road network, removing bottlenecks through minor construction work and improving traffic management, rather than expanding capacity. The earlier *Report on Transport and the Environment* published in 1994 defended 'placing significant constraints on the future evolution of the transport system', arguing that 'transport involves large costs, some incurred directly or indirectly by users, and some as a result of its environmental effects. Hitherto, most of the latter costs have fallen on the community rather than on the users or the builders of the transport system. Seriously misleading price signals have resulted, leading to decisions in all areas of transport which have harmed the community.' (*Report*, 1994, §14.5).

To an economist the argument for restraining traffic growth is that the marginal social cost (MSC) exceeds the marginal social benefit (MSB), which for road users is normally equal to the marginal private benefit, which at the margin will equal to the private cost of travel (including the road taxes). Social costs arise from two main sources - environmental pollution and congestion costs. Current road taxes are sufficient to cover the social costs of pollution for most vehicles (Newbery, 1998a), but congestion costs can greatly exceed private costs and vary considerably across roads in different places and at different times. Road pricing offers the prospect of charging motorists more where congestion costs are high. The *White Paper* proposes legislation to enable pilot schemes for road user charging to be developed. Schemes may be developed, for example, to help to meet transport and environmental objectives in urban or rural areas, or on bottlenecks on specific roads or at certain times of the day or year. (§4.100).

The Theory of Road Pricing

The design and benefits of road pricing are illustrated in Figure 3, which illustrates a simple model of congestion[3]. The bottom curve AEB shows the average social cost per km (ASC) of a vehicle as it varies with traffic flow. The ASC includes all the operating costs of the vehicle, the external environmental costs (of pollution, noise, CO , etc), and the cost of the time of the occupants, which increases with traffic flow as congestion causes speeds to fall and time taken per km to rise.

The vehicle pays fuel taxes, part of which covers the costs of environmental externalities, and part of which covers the cost of providing road space. Fuel taxes may produce more revenue than justified by these costs, and these may be justified in part as rationing under-provided road space, and/or raising additional revenue for the Government. Define the road charge element to be the fuel tax *less* the costs of the environmental externalities, and add this road charge part to the ASC to give the APC, which is shown as the line parallel to the ASC, just above it through C (where the road charge element is 3.21p/km, shown as CB).

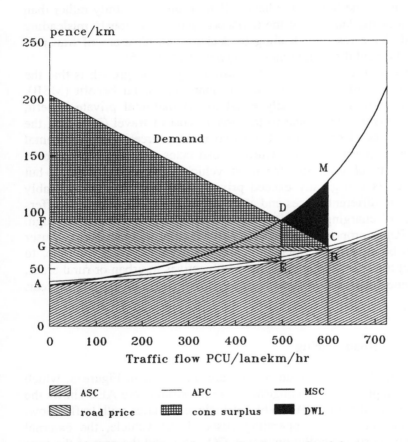

Figure 3 Benefits of travel reduction by road pricing

The demand for travel measures the willingness to pay for trips, converted here into a willingness to pay per km. Higher prices result in fewer trips and hence a lower traffic flow. Demand here is shown as linear with an elasticity at the equilibrium C of -0.5. In equilibrium, motorists face the tax-inclusive cost of making trips and choose a level of demand accordingly, so the equilibrium flow is at C (at 600 PCU/hr).

The *marginal* social cost or MSC, is, however, higher than the ASC because each extra vehicle adds to the flow, reducing the speed of other vehicles, and raising their cost. The curve ADM shows the MSC, and it intersects the demand schedule, which measures the marginal social benefit (assuming that there are no other unaccounted for beneficiaries), at D (about 500 veh/hr), which corresponds to the efficient level of travel demand. In this case, then, demand should be reduced (by about 100 veh/hr).

One way to ration the road space efficiently would be to use road pricing, that is, to set a price ED (at a level of 35.7p/PCU km) to use the road (at this time and place, given the level of traffic demand and road capacity). The revenue from the road price is also shown cross-hatched in Figure 3, as is the original road charge (which goes out to CB).

The cost of the inefficiency of charging CB instead of DE is measured by the black triangle DMC, which is the dead weight loss of inefficient pricing. The concept of a dead weight loss, familiar to economists and much used to analyse the efficiency of tax systems, is not at all familiar to non-economists, and warrants explanation. The area under the demand curve but above the price line, shown hatched, measures the extra amount that motorists would be willing to pay for trips, compared with what they actually have to pay. It is the consumer surplus, or the unpaid-for consumer benefit of access to the road system at the specified price. The revenue from road prices and charges, also cross-hatched, represents transfers from consumers to the Government which can be used for the benefit of the community (including road users). But beyond D, the marginal social cost exceeds the marginal social benefit, and the excess cost is measured by the difference between them. The total excess cost is thus the triangle DMC, shown in black. If a road price increased the total charge to DE, the consumer surplus would fall by DFGC, but revenue would increase by this amount plus DMC, so society as a whole would gain the area DMC that was previously lost. If the cost of introducing road pricing is less than this gain, then road pricing looks desirable on simple cost-benefit grounds. The benefit of reducing travel demand would be the triangle DMC, less any implementation costs.

Of course, one can go further than this and quantify who gains and loses by the policy. That will depend on whether the Government decides to reduce fuel taxes when it introduces road prices. If so, then some other less congested journeys will become cheaper and it is possible that the majority of road users finish up better off than before. The Government will presumably collect more revenue (some of which may be offset by reduced fuel taxes and the costs of introducing the scheme), while others may also benefit - those using public transport, whose journey times improve, those experiencing the pollution and noise, which should fall, etc. If fuel taxes are not reduced, then road users as a class will lose, and tax payers will gain from the extra tax and charging revenue.

The repercussions of road pricing are likely to be wider, however. Shop-keepers may sell less or more, depending on a variety of factors (such as whether other towns that are less congested are now more attractive places to visit and shop, or this town now attracts higher income shoppers who find the ambience improved), while the value of land in different places will also change in response to changes in the attractiveness of the town and the cost of travel. In due course there may be longer run responses to all these changes, as people and businesses relocate in response to changed costs and prices, and these longer run effects may be considerably larger and more important than the immediate effects.

The Case for Motorway Tolls

The case for road pricing in urban areas rests on the claim that the marginal social cost of congestion is high and considerably exceeds the current level of fuel tax. This is reasonably uncontentious, though it needs further discussion, which is deferred until later in this paper. But the section of the *White Paper* that discusses road pricing is headed 'charging users on motorways and trunk roads', where the case is nothing like as well based. Two points need to be made here. The first, and most important, is that congested inter-urban roads are sufficiently cheap to expand that current road taxes more than cover the cost of expansion per veh.km - often by a factor of two. The second is that current Government thinking is well summarised in its Road Traffic Reduction (National Targets) Bill, where the aim seems to be to cut planned inter-urban road construction as a mechanism for restraining the growth in road transport demand.

Figure 4 shows the effect of a policy of restraining demand by restricting road supply. It shows the effect of reducing the amount of road capacity by one-third, while leaving demand and road taxes unaltered (which has a similar effect to not expanding the road network in line with demand - the much maligned 'predict and provide'). The figure graphs the APC before (APC0) and after (APC1) the capacity reduction, and the associated MSC. The area under the MSC is the total social cost of meeting the level of demand, and is initially given by the left-hatched area AEFD. After the capacity reduction congestion worsens, and travel demand is curtailed by the increased private cost of travel (APC1 is considerably higher than APC0), so the new equilibrium involves a reduction in travel from 600 veh/hr to 500 veh/hr - a similar impact to the efficient road pricing shown in Figure 3.

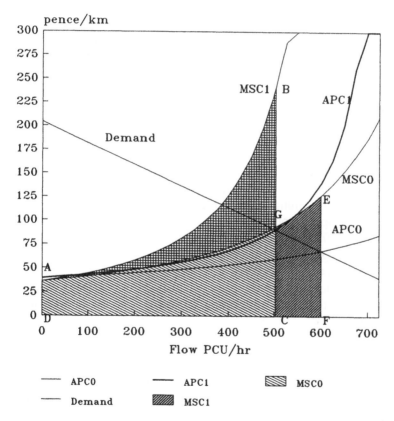

Figure 4 Costs of reducing capacity to reduce travel demand

However, rationing by congestion is extremely inefficient, and the social cost of providing this level of travel is given by the right-hatched area ABCD. The increase in cost is area ABG *less* area GCFE. In this example, the total cost per hour of meeting travel demand before reducing capacity is £389 and after is £438, so costs have risen by £49/hour, an increase of 13% despite a fall in traffic demand met of 17%. In addition, consumer surplus has fallen as the perceived private cost has risen, and the resulting loss of consumer surplus is £123/hr, making a total cost of reducing road capacity of £172/hr, or 44% of original total cost.

The inefficiency of the vehicle transport system has deteriorated by the same amount, measured by the increase in the size of the dead weight loss triangle (not shown, but readily visualised by analogy with Figure 3). Rationing demand by restricting road space is extremely inefficient, and the Government is therefore to be complimented for choosing the less inefficient solution of restricting demand by road pricing. It is to be criticised for creating an avoidable inefficiency in raising the price above the marginal cost of supply of road space. The Government is therefore exercising monopoly power in restricting supply to drive up scarcity prices for revenue raising purposes.

Lest this claim seem unreasonable let me provide some evidence of how matters are managed on the Continent. Figure 5 compares congestion in the UK with other European countries, and it is striking that compared to France, Britain has more than five times as much traffic delay, while projections in the Department of Transport suggest that the situation will rapidly deteriorate in the coming years. Figure 6 compares road finances in Britain and other European countries, and shows the road tax revenue per kilometre of road for 1996, and also for Britain for 1998. Although in 1996 the Netherlands had higher petrol taxes per kilometre of road, by 1998 the UK had exceeded even the Netherlands.

Thus other countries are better able to provide adequate transport infrastructure and avoid serious congestion problems, and do so with less revenue per vehicle km. The Netherlands, which is considerably more densely populated that the UK, has a completely different policy to expanding capacity on inter-urban roads and motorways to alleviate congestion. Their cost-benefit analysis shows that once a motorway is congested more than 3% of the time, then capacity should be increased. France manages to accommodate a far larger volume of traffic around the Paris conurbation with far less congestion than the M25. By most indicators British inter-urban road transport lags behind Europe, and the Government appears to be exercising monopoly power rather than acting as a public service provider.

To conclude this section, an efficient road transport policy would rarely need to resort to road pricing on inter-urban roads and motorways. Road fuel taxes are a simple and cheap-to-collect method of charging motorists

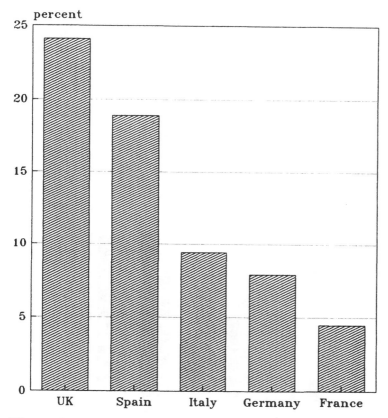

Figure 5 Percentage of congested road links (delay of one hour or more per day)

Source: The State of European Infrastructure, European Centre for Infrastructure Studies

for distance, and if set at the right (rather lower) level, will finance expansions as they become cost-effective. Once the predicted traffic increase generates enough revenue to finance an additional lane then expansion is justified[4]. Most inter-urban roads experience moderately constant traffic flows in daylight hours, but peri-urban motorways may experience pronounced peak flows, and these may benefit from some form of road pricing. One attractive solution is that adopted in California, where a lane for high occupancy vehicles is also made available to toll-paying cars. This has the considerable political appeal that those who do not wish to pay the toll can choose to travel in the congested lanes, while the toll ensures that the tolled lane will be less congested. In its most sophisticated form, the route operator attempts to guarantee a transit time by varying the toll, though granting the right to the entire trip at the initially announced toll for each vehicle that enters the paying lane. As traffic builds up, the price is raised to maintain a constant flow rate into the tolled lane.

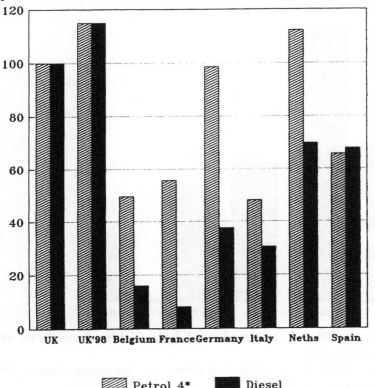

Figure 6 Road tax revenue per km road - 1996
(Index numbers UK 1996 = 100)
Source: Tax/litre x VKT/km derived from Transport Statistics GB 1996

The critical question to ask of introducing tolls on such motorways is whether the *reduction in dead weight loss* outweighs the costs of installing and operating the system by enough to overcome the political ill-will created by the extra charges. Unfortunately, the normal question asked is whether the *revenue* covers the cost and makes a profit for the operator, who, in this era of PFI, is likely to be a profit maximising private company, not a social welfare maximising supplier of a public service.

Road Pricing in Urban Areas

The research Georgina Santos and I have been conducting under our ESRC grant aims to quantify the dead weight losses associated with congestion in the UK, and as congestion is primarily an urban phenomenon we have started with an example of the convenient and highly congested city of Cambridge. Earlier attempts to estimate the appropriate level of road prices in urban areas (Newbery, 1988, 1990) were based on speed-flow relationships derived from floating car measurements and embodied in the COBA manuals of the Department of Transport. The derivation of the road price then consists of computing the externality associated with congestion. If $c(q)$ is the vehicle cost per km when traffic flow is q, and $C = cq$ is total cost per km of the traffic flow, then the marginal social cost of traffic is dC/dq (shown as the MSC in Figure 4) while the private cost (APC) is $c + \tau$, where τ is the road tax per km. The value of c is in turn found from the distance-related costs a p/km and the time-related costs (essentially the value of the time of the occupants) b p/hr, when the speed is v km/hr:

1. $$c = a + b/v .$$

The COBA speed-flow relationship is given by

2. $$v = v_0 - \beta q ,$$

where $\beta = 0.035$ approximately in urban areas. The congestion externality, e, per VKT is the marginal social cost, dC/dq, less the private (pre-tax) cost, c, and by differentiation of (1) and (2) is

3. $$e \equiv \frac{d}{dq}(cq) - c = q\frac{dc}{dq} = \frac{b\beta q}{v^2} = \frac{b(v_0 - v)}{v^2}.$$

Given observations on q and v (or just v, as in the last expression, and relying on 2. to estimate the other), one can then compute the optimal toll $p = e - \tau *$, (where $\tau *$ is that part of the road tax left over after dealing with other environmental externalities). One can also compute the dead weight loss by simple geometry from the approximate formula

4. $$L = \frac{1}{2} \left(\frac{(e - \tau *)^2}{\dfrac{c + \tau}{\varepsilon} + \dfrac{2 e v_0}{v}} \right),$$

where ε is the elasticity of demand at the point $(c + \tau, q)$, and L is the dead weight loss per VKT[5].

This approach assumes that the speed-flow relationships derived from the observations accurately describe the causal features of congestion in towns. The main reason for doubting this is that most delays take place at intersections, while the speed-flow relationship is normally thought of as applying on a link. In our research we have therefore taken a different approach, deriving estimates of trip times and traffic costs from the SATURN (Simulation and Assignment of Traffic in Urban Road Networks) model. SATURN is a software package that simulates and assigns traffic in urban road networks and iterates until the equilibrium is reached. Equilibrium is a situation in which no trip maker can reduce his or her generalised cost, defined as the cost of time plus the cost of distance. The model stops iterating when the cost of travel on all routes used between each O-D pair is equal to the minimum cost of travel and all unused routes have equal or greater cost.

We do indeed find that almost all (simulated) congestion takes place at junctions, and unfortunately there is no simple relationship as in 2. that allows an algebraic derivation of externalities. One obvious difference between 2. and the junction congestion is that it is flows on *other* arms that delays traffic on a given entry arm, so that there is no longer the pleasing symmetry of vehicles delaying each other equally. We are currently using SATURN to see if we can derive rough graphical relationships between flows on arms and delays to traffic, and it may be that the COBA speed flow relationship (where the constants depend on the number of junctions per km, amongst other factors), can be recovered for a 'typical' urban configuration (which is what we are seeking).

The other implication of the view that congestion is a junction phenomenon is that there is an important distinction between short and long run marginal social costs which has implications for road pricing. Consider the evolution of traffic growth in a town like Cambridge. Initially there are few roundabouts and traffic lights, and mostly junctions are not too congested. As traffic increases, it becomes necessary to improve flow at junctions, and where these are upgraded with traffic lights, delays now occur on the priority routes where previously they did not, though to the (greater) benefit of those on minor roads. The effect is that in uncongested conditions, average speeds will be lower than before, but in congested conditions, flow will be improved and speeds higher than they otherwise would have been.

Investment in increasing peak capacity (in traffic lights, etc) thus shifts the location of the average and marginal social cost curves, tending to raise the ASC but reduce the difference between the MSC and ASC. In an extreme situation the traffic lights will be set to send packets of traffic through the network at moderately constant speeds, so the effect of an extra vehicle will either be zero (if the packet has 'space' for an extra vehicle) or to delay following packets until a space appears. The other effect (which is hard to capture on simulation models) is that more traffic at peak hours means that a higher fraction of the network reaches criticality, so that a random disturbance (parked vehicles, buses stopping for passengers, accidents, breakdown, traffic light failures, etc) will cause a far larger effect as other parts of the network are disrupted. Recovery times to restore normal flow will also be longer at high traffic levels. More traffic in the off-peak period is less likely to be so disturbed, and the effect of extra vehicles may be negligible. The average social cost curve may thus approach a reverse-L shape as traffic management becomes more intense.

What are the implications of these observations? First, it may be dangerous to use urban speed-flow relationships without validating them against more detailed traffic flow models. Secondly, that task is complicated because standard traffic assignment models do not capture the random disturbances that may greatly increase the congestive effect of extra vehicles joining the same traffic flow. Thirdly, if the system of traffic management is designed for peak flows, there may be relatively low congestion in the off-peak periods, even though the speeds may be low, suggesting congested conditions. Effectively, the term v_0 in 2. and 3. is reduced by the traffic management schemes.

If we are considering a system of road pricing to reflect the social costs of congestion and give better signals for efficient use of road space and location decisions for residents, shops and businesses, then we need a soundly based attribution of costs caused by actions. The same problem arises in other network industries such as setting charges for using the National Transmission System (NTS) for gas and for the National Grid in the case of electricity. The National Grid loads all charges on to the peak three half-hours separated by ten days, while in the NTS shippers pay a capacity charge which gives them the right to demand up to that capacity on any day, including the peak day. The fraction of total costs paid as a capacity rather than commodity charge has been raised from 50% to 60% and may be raised further, perhaps to as much as 90%. In both cases the costs recovered by these peak charges are the investment costs of providing capacity and the operating costs of the system. It is not thought necessary to adopt market clearing scarcity prices, for the security margins covered by these charges rarely make such rationing necessary (though gas shippers can select interruptible tariffs and avoid paying capacity transmission charges), and the charges are adequate to finance expansions of the system to maintain the security margins.

Urban roads are different in that their capacity cannot readily be increased, though traffic management measures do this to some extent and are costly, not only in the direct investment cost sense, but because they raise costs to motorists in the off-peak. On the analogy with pricing other networks, one would argue that all the costs of providing capacity on the urban road network, including these investment costs and the associated extra off-peak costs, should be charged to motorists using roads at the peak, since it is the peak flows that give rise to the extra costs visited on the off-peak users. Standing back somewhat, the object of road pricing (and other measures) would be to encourage towns to reach an optimum size and degree of traffic intensity, so that the total social surplus across towns of different size and location (benefits of whatever it is the towns offer less the full costs of access) is maximised. It is not clear whether this can be achieved just by road pricing, or whether additional instruments are needed. At present land-use is rationed by planning controls, and there have been proposals to supplement these by market-based instruments such as taxes on parking spaces. It may be that efficient urban road pricing would make these additional taxes unnecessary. As urban road capacity is very expensive to expand, the marginal cost of providing extra capacity is likely to be very high, while the road price needed to ration the scarce space efficiently in peak hours is also likely to be high. If capacity is optimally provided, then the marginal cost of capacity and the efficient road price to ration that capacity would be equal. That is, the revenue collected from the extra vehicles allowed by an increase in capacity would just pay for the extra cost of providing that extra capacity.

The normal reason for employing additional taxes (parking taxes, etc) is that otherwise there may be a shortfall between efficient prices and costs. It seems more likely that efficient road prices would produce a surplus, because the marginal cost of urban expansion exceeds the average cost of the urban road network, so additional taxes would seem unnecessary. The other reason for using additional taxes and other instruments is that it may not be possible to set the road price at the efficient level in all conditions, and additional instruments may then improve the final outcome.

Results of our ESRC Research on Congestion

At present our research is incomplete, and we are not in a position to provide satisfactory answers to all these questions. We do have some preliminary results for York and Cambridge, which are reported more fully in Santos and Newbery (1998). The York model is more suitable to the present purpose as it is confined to the city within the outer ring, and treats all journeys as starting just outside this ring[6]. The Cambridge model includes a very extensive outer ring of relatively uncongested roads. Thus the average recorded trip length in the York model is 6.5km, while in Cambridge it is 20km. In Santos and Newbery (1998) we report the results of breaking the Cambridge model into an outer and inner zone, and we find that the inner zone, of about the same area and traffic volume as York, yields similar results. We shall therefore take the York model as a suitable example of a congested urban area (see Table 2).

Our method of measuring the MSC is to compute the total costs at various levels of trips, and then compute the incremental cost per trip, expressed per km travelled. One interesting finding is that as the number of trips increases, so vehicles choose less congested routes and travel longer distances per trip. The average cost per km could even fall, but the cost per trip rises, and it is this increase in costs that is caused by additional traffic (measured by trips demanded). Concentrating on simple speed-flow relations along a link of fixed length would not pick up this system-wide effect, and would understate the social costs of extra traffic.

Using an initial set of values (for time of 14.4p/min and for distance of 13.66 p/km), the average private cost (including fuel tax of 4.2p/km) APC, marginal social cost MSC (including environmental externalities of 1p/km), and the marginal congestion cost MCC ≡ MSC-APC, which is the natural choice for a road price, were then computed[7]. The initial results are shown in Table 3 for York:

Table 2   Traffic data for York model

Time of day	Number of trips	Veh-km per hour	Veh-km per veh
Morning peak	33,848	218,500	6.5
Evening peak	33,754	216,130	6.4
Off peak	24,074	151,390	6.2

Source: York City Council and calculations by G Santos made using
SATURN. Morning peak: 8-9am; afternoon peak 5-6pm; off-
peak average over 14 off-peak hours

Table 3   Congestion externalities for York

	Morning Peak	Evening Peak	Off peak
APC in p/km	38.2	38.8	35.2
MSC in p/km	74.5	78.6	41.2
MCC in p/km	36.3	39.8	6.0

Source: Calculations from SATURN output

These estimates of congestion externalities are consistent with those
derived from urban speed-flow relationships, of the kind published in
Newbery (1990). That showed MCC of 36p/km in urban centres at the
peak, and 16p/km outside the central area also at the peak (in 1990 prices).
The ratio of VKT travelled in non-central areas is about three times that of
central areas, so the weighted average MCC would be 21p/km at 1990
prices or 28p/km at 1996/7 prices. Santos and Newbery (1998) give
comparable figures for the centre of Cambridge. Note that for York the
off-peak traffic flows are measured for 11am-noon, and these will be
considerably higher than at other off-peak hours. In later work we have
estimated the costs at each hour of the off-peak period and then computed a
weighted average that can be used to estimate the daily congestion costs.
The results are quite similar to those reported here.

Our early findings about the size of the dead weight loss suggest that it
is small compared to the revenue that would be collected from an optimal
road price. The required road price is somewhat less than the estimated

congestion externality, as demand would fall were road pricing to be introduced, but the difference is modest and of the order of a few percentage points of the congestion externality. The ratio of the dead weight loss to revenue is about 4% averaged over all the off-peak hours (assuming in each hour the optimal road price could be charged), rising to 10% at peak hours, though the ratio rises quite rapidly as traffic increases above the calibrated level of traffic. These rather low estimates of dead weight loss suggest that the benefits of road pricing could be rather small compared to the cost of introducing and collecting it, and the main attraction of road pricing is likely to be the revenue that it generates. There is an obvious danger that road pricing will become just another tax, and possibly one that is more expensive (in accounting and political terms) than alternatives.

Figures 7 and 8 give a visual account of the York results. Figure 7 plots the average trip speed as a function of total VKT in the York model zone, and shows the close linear relationship that is found in simple link speed-flow relationships. Indeed, if the number of VKT is divided by 450, the slope would be the same value of 0.035. One way to interpret that is that York behaves as though it has 450 equivalent lane km of road capacity. This coincidence between the link and network speed-flow relationships may indeed just be a coincidence as early results from Cambridge show a far from linear relationship, so it will be important to explore this relationship further using other towns. Figure 8 shows the APC and two estimates of the MSC, one based on smoothing the APC by taking a 3-pointed moving average, the other, bold, taking a 5-pointed moving average, and then computing the smoothed total social costs (deducting fuel taxes and adding 1p/km for pollution) from which the MSC was computed. The demand curve through the actual morning traffic flows for which the model was calibrated is shown, with an elasticity of -0.2.

The MSC becomes quite irregular at higher traffic flows. Part of the reason is that as congestion worsens, cars take different routes and typically make longer journeys. Figure 7 shows the average journey length per trip rising from just under 6 km to just over 7 km (admittedly over a huge range in traffic flows). Route switching at threshold values may account for the apparent unevenness in the MSC schedule. Otherwise the shape of Figure 8 is similar to that of Figure 3 which was derived from the standard speed-flow relationships. Whether this will be found in other congested urban environments awaits further research, but if so, then perhaps the traditional way of computing congestion tolls is not too misleading.

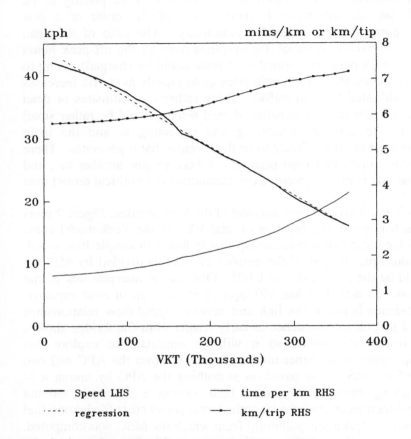

Figure 7 York SATURN model. Speed vs traffic volume

The Design of Road Pricing Systems

The next question would be the design of the system of road pricing, which includes both the choice of technology and the formula for setting the prices (taking as read the need for a regulatory authority to limit the total

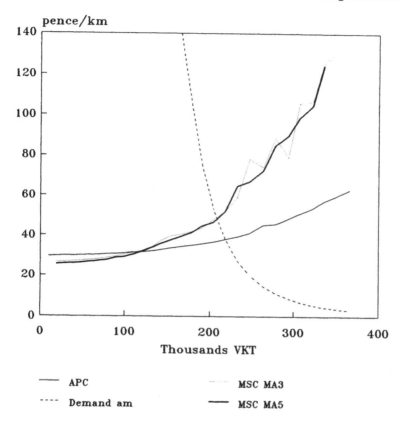

Figure 8 York SATURN model. Private and social costs

revenue that can be raised by this process). The main criterion for the choice of technology is that it should be cheap, work throughout the UK (and possibly in other EU countries), be reliable, and tamper-proof. At some stage presumably a standard will be agreed and the design can be stabilised. The choice of price-setting formulas that have been considered include cordon tolls (which might just charge for access to the whole central zone, or for crossing between sub-zones), distance tolls within areas, time tolls within areas, or some combination of the last two. Assuming they are all equally feasible and costly (which is most unlikely), how do they compare?

A cordon toll penalises short journeys relative to longer journeys, and raises obvious problems for those who live inside the cordon if the intention is to charge differentially for the peak and off-peak. It can have undesirable impacts on business location, encouraging non-central annular journeys which are both inefficient and lead to congestion that is harder to alleviate by public transport. It has the obvious attraction that if vehicles are to be observed and then billed, fewer gantries are required.

The alternative is a vehicle meter which is charged up (perhaps with a smart card) and deducts for travel and possibly also for parking, once activated by a road-side transmitter. This may require more investment in the vehicle device, and may be harder to make tamper-proof (though this might be overcome by interrogating it from the roadside to test its status), but it allows vehicles to be charged by distance and/or time.

The Cambridge road pricing experiment was an ingenious attempt to charge for congestion, on the principle that if the vehicle was travelling at less than a pre-set speed, it would then be in congested conditions, and should pay per unit of time so congested. It ran into two problems - most delays are caused by junctions and by traffic on *other* roads, creating a sense of injustice, which exacerbates the second problem that it discourages safe driving where this involves slowing down. The same argument applies to time-based charging, where most motorists are sufficiently impatient that charging them more for a poorer quality of service is likely to make road manners worse rather than better. Psychologically, then, distance-based charges at peak hours, perhaps at different rates in different zones, have the merit of predictability (which matters for encouraging efficient route and location choices), fairness (being comparable to bus and taxi charges which are distance-based, and in proportion to the service provided and congestion caused), and safety.

Problems to Address

Apart from the very considerable design problems (which smarter electronics will doubtless assist) there is the obvious cost of installing the system in some 22 million vehicles. If the schemes are confined to a limited number of towns and vehicles, then there is the problem of dealing with occasional visitors (perhaps to be dealt with by issuing either a daily permit for entering the town, or renting a device which can be returned). The more serious problem is to determine what fraction of revenue is to be collected from road prices and road taxes. If the road pricing is additional to existing and escalating road taxes, they will be more vehemently opposed than if there is a deal to cap the total payments from vehicles, to be apportioned between road prices and residual charges based on fuel (as a simple proxy for distance and one with very low evasion and low collection costs, particularly suitable for inter-urban roads) and VED. The problem with this approach is that until all vehicles are equipped to pay road prices, some may be overcharged relative to others. It might be possible to handle this by rebating VED to those motorists installing meters and paying road prices, or by giving them a credit on total km driven per year (which can presumably be reliably measured by the same meter) as an offset against that part of fuel taxes paid that would eventually be replaced by road pricing.

The other problems have already been mentioned - designing a system of dual funding of transport improvements that delivers value for money, commands confidence, is accountable, and subject to regulatory oversight. There are particular difficulties in ensuring that investments in improving public transport (mostly privately owned) satisfy the stringent cost-benefit criteria that road investments will be required to meet, and there is the pervasive concern that giving authorities the power to raise revenue without adequate regulatory scrutiny will encourage them to restrict supply monopolistically and to reduce quality.

Conclusions

Road pricing is a welcome addition to the management of the road transport system, and making this possible may be the major legacy of the *White Paper*. It raises a host of issues none of which has been adequately addressed in the *White Paper*, and some of which have been carefully evaded. There is no satisfactory resolution of funding mechanisms, and a worrying reliance on expensive PFI. There is no mention of the need for regulatory oversight, and no discussion of the relationship between the two methods of charging road users. The best hope is that the process of drafting legislation for road pricing might require these issues to be addressed, and that the research to be commissioned, along with the pilot studies, will give a more solid basis to urban road pricing than we presently possess. The other hope is that the Government might at last realise that tolling inter-urban roads is an admission of an inability to provide an adequate core network, which would be sensibly financed by fuel taxes or charges as at present.

Appendix: Extracts from A New Deal for Transport: Better Transport for Everyone

Funding Transport

4.4 This is the first comprehensive White Paper on transport policy for 20 years. But it is not the end of the story: we need to continue to work on our policies and not wait another generation before we take stock of how we are getting on. To help keep the debate alive and to continue building on the consensus, we will establish a new independent body - the Commission for Integrated Transport (CfIT) - to provide independent advice to Government on the implementation of integrated transport policy, to monitor developments across transport, environment, health and other sectors and to review progress towards meeting our objectives. Its remit will include:

- reviewing and monitoring progress towards objectives and targets set out in the White Paper;
- continuing and refreshing the transport policy debate;
- fostering consensus among practical providers;
- identifying and disseminating examples of best practice from home and abroad;
- advising on developments in Europe, including relevant EU initiatives;
- advising on the role of existing and emerging technologies.

We will ask the Commission to advise us, among other things, on:

- setting national road traffic and public transport targets;
- the revisions we will be making to the 1997 National Road Traffic forecasts in the light of the measures in this White Paper;
- lorry weights and the development of rail freight;
- the review of transport safety arrangements;
- progress on the take-up of green transport plans;
- the new rural bus partnership fund in England;
- how to secure best value from public subsidy for the bus industry in the longer term;
- public expenditure priorities for integrated transport in the longer term;

- research, in particular with a view to gaining a better understanding of the costs and benefits of transport and how these relate to the costs faced by users.

4.5 Our new approach to transport is not the property of any one party or interest group. The Commission will have an independent chair and a small permanent core of members, chosen in part to represent particular interests but principally for their expertise and impartiality. It will include representatives of Scotland, Wales and Northern Ireland, someone from the science and technology community and a transport user representative. The Commission will be required to consult widely with providers and regulators, central and local Government, regional bodies, interest groups, trade unions, business and users. It will also draw on the expertise and resources of other organisations and individuals drawn in for work on particular topics. The arrangements for dealing with devolved matters will be set out in the Scottish Integrated Transport White Paper and the Welsh transport policy statement.

4.6 The Commission will make recommendations to Ministers and prepare an annual report on the implementation of the new approach, including progress towards meeting targets, the impacts of key policy initiatives and priorities for further action.

4.7 Transport makes a significant call on the public purse - this year, for example, planned expenditure includes some £1.6 billion on railways in Great Britain, around £3 billion on local transport in England and £1.3 billion for the English trunk and motorway network. We will ensure that public expenditure on transport is firmly directed towards delivering the New Deal for transport. In addition, through partnership with the private sector, we expect to see the level of privately-financed investment in transport increase by at least a half over the next three years.

4.8 ...Our Economic and Fiscal Strategy sets out our framework for future spending which will allow real current spending to grow in line with the growth of the economy, whilst enabling us to increase capital spending to double the level of net public investment as a share of GDP. Our transport infrastructure in particular will benefit from this significant boost to public investment.

4.9 But we have to determine the balance of expenditure on public services according to our economic and social priorities. Although transport investment at national level could be funded by dedicating particular streams of taxation income, as some suggest, that would inevitably restrict our ability to use that income flexibly both for transport and for other priorities such as education and health.

4.11...Recognising this, we announced in the Economic and Fiscal Strategy a major reform of the public spending rules. We have abolished the annual spending round which encouraged short-termism and inefficiency. Firm spending limits for the next three years will give us greater certainty and stability to plan and manage our programmes sensibly.

Tackling Congestion and Pollution on Local Roads

4.94 But experience has shown that improving public transport and related traffic management measures whilst necessary are not sufficient in many cases. We will therefore introduce legislation to allow local authorities to charge road users so as to reduce congestion, as part of a package of measures in a *local transport plan* that would include improving public transport. The use of revenues to benefit transport serving the area where charges apply, which in many cases will mean supporting projects in more than one local authority area, will be critical to the success of such schemes.

4.95 Carefully designed schemes should reduce traffic mileage and emissions, bringing significant improvements in air quality, reducing noise and greenhouse gas emissions and relieving congestion. This will benefit pedestrians, cyclists and public transport, including more reliable and quicker bus services and more reliable delivery times for freight. Less congestion also means shorter and more reliable journey times for those who continue to drive. Charging will provide a guaranteed income stream to improve transport and support the renaissance of our towns and cities. The availability of a revenue stream will also open up the scope for greater involvement of the private sector working in partnership with local authorities.

4.96 In rural areas, road user charging is most likely to be used where there are significant problems caused by very high levels of seasonal traffic, for example, in tourist areas such as the National Parks. We would welcome proposals for such initiatives to provide the basis for pilot schemes in rural areas.

4.97 Primary legislation will be needed. Subject to that being in place, we will then work with local authorities and other interested organisations on a number of pilot schemes individually approved by the Secretary of State (in Scotland, by the Scottish Executive). The effects of these schemes will be monitored and used to inform the design of future schemes.

4.98 We will issue a consultation document with proposals for how road user charging schemes should operate. This will deal with different ways of implementing charges: electronic schemes, schemes where drivers must buy and display a permit and schemes using toll booths. It will seek views on how best to ensure the active involvement of local people, business and others in the development of schemes so that proposals attract public support. We will also be seeking views on how such policies will impact on the mobility of disabled people.

Charging Users on Motorways and Trunk Roads

4.100 Our proposals for legislation to allow road user charging will enable pilot schemes to be developed in a variety of circumstances. Schemes may be developed, for example, to help to meet transport and environmental objectives in urban or rural areas, or on bottlenecks on specific roads or at certain times of the day or year. Such schemes may also be developed on trunk roads and motorways, either on a self-standing basis or as joint schemes with local authorities. Pilot charging schemes will be individually developed and designed to take into account the local transport network, ensuring in particular that unacceptable diversion does not take place on to local roads. We will also consider for each scheme how best revenues generated may be used to provide related benefits locally which might otherwise be unaffordable, including better means of securing the environmental acceptability of transport infrastructure.

References

Newbery, D.M.(1988) Road User Charges in Britain *Economic Journal* (Conference), 161-176.

Newbery, D.M.(1990) Pricing and Congestion: Economic Principles Relevant to Pricing Roads *Oxford Review of Economic Policy*, Vol 6 (2), 22-38, reprinted as ch. 13 in Layard, R. and Glaister, S. (eds.) (1994) *Cost-Benefit Analysis* (2nd ed) CUP, Cambridge.

Newbery, D.M.(1994) The Case for a Public Road Authority *Journal of Transport Economics and Policy*, XXVIII (3), September, pp.325-54; also in *Minutes of Evidence*, Transport Committee Fifth Report *Charging for the Use of Motorways*, Vol II, 74-87 HC 376-(II).

Newbery, D.M. (1998a) *Fair Payment from Road-Users: A Review of the Evidence on Social and Environmental Costs* Report published by Automobile Association, February.

Newbery, D.M. (1998b) Fair and Efficient Pricing and the Finance of Roads (The 53rd Spurrier Lecture), *CIT Proceedings* (forthcoming) (available on my web site as the 53rd Spurrier Lecture).

Santos, G. and Newbery, D.M. (1998) Social Costs of Congestion in Cambridge and York, mimeo Department of Applied Economics, Cambridge.

Notes

1 Support from the ESRC under Grant R000222352 *Quantifying the Costs of Congestion*, being conducted with Georgina Santos, is gratefully acknowledged.
2 The time profile of the price cap may need adjusting to ensure that any investment can be readily financed out of cash flow and borrowing without prejudicing prudent financial requirements, while ensuring that the present value of the revenue stream is appropriate for the opening asset value and projected flow of operating expenditure, but this does not affect the principle of setting the level of the price cap.
3 The model is calibrated to 1996 prices using the COBA speed-flow relations for central urban areas, for which the speed flow relationship is $v = 33.875-.03q$ kph, where q is PCU/lane km/hr (and PCU is Passenger Car Units). The value of time is £8.64/PCU hr, the externality is 1p/PCUkm (from Newbery, 1998a), the fuel tax is 4.21p/PCU km, so the road charge element (tax less externality) is 3.21p/PCU km. The vehicle cost per km excluding the value of time and road charge element is 9.45p/PCU km.
4 There are some slightly delicate issues of timing the lumpy expansion, which should take place when the present value of social benefit less cost is maximized, but the principle is essentially the same.
5 If the slope of the MSC at point M in Figure 3 is θ, and the slope of the inverse demand curve, $P = A - Bq$ is B, and if the horizontal width of the triangle DMC in Figure 3 is Δ (ie the distance from D to the perpendicular MC), then $\Delta B + \Delta\theta = p$ (= CM in Figure 3), and the area of DMC is $\frac{1}{2} \Delta e = \frac{1}{2} p^2/(B+\theta)$. If the elasticity of the linear demand curve at point C is ε, then $B = (c+\tau)/\varepsilon q$. Finding θ algebraically requires maintaining the hypothesis of equation (2), in which case $\theta = dS/dq$, where S is the MSC or the curve ADM in Figure 3, $S = c + qdc/dq$, so $dS/dq = 2dc/dq + qd^2c/dq^2 = 2b\beta(v+\beta q)/v = 2ev_o/(qv)$. The dead weight loss per VKT requires dividing the area DMC by the level of VKT, q.
6 The zone modelled is delimited by the A64 and A1237 and is roughly 11 km N-S and 14 km E-W.
7 These values have subsequently been refined to distinguish between peak and off-peak costs, and to use more reliable estimates of fuel consumption by class of vehicle. See Santos and Newbery (1998). In early November we were informed that there is a bug in the current issue of the SATURN program which may affect the reliability of the estimates presented here. They should therefore be treated with some caution and not cited. We hope to prepare revised estimates when the nature of the problem has been resolved and issue them in Santos and Newbery (1998).

5 Pricing of Infrastructures

DAVID STARKIE

Charging for Access to Rail Track

An efficient structure of rail access charges should provide appropriate incentives for the effective management of the existing network and provide adequate financial resources to allow for its enhancement when this is justified by market demand. It is by reference to this benchmark that Chris Bolt, in his contribution, reviews the EC 1998 proposals for charging for infrastructure use and concludes that the Commission's proposals have significant limitations. It is also evident from his comments that the current structure of Railtrack's access charges, established at the time the privatised railway was put in place, is also unsatisfactory. This current regime is seriously deficient and this reflects in part the pace at which the railways in Great Britain[1] were reorganised in the mid-1990s. The EC proposal, notwithstanding Bolt's reservations, would in my view do much to eradicate these deficiencies.

The basic problem with the EC proposal is considered to be its sharp focus on short run marginal cost pricing (SRMC) which is defined to include 'congestion and environmental costs but not capital costs'. However, I do not see a fundamental problem with the EC proposals and its focus on SRMC. From a theoretical standpoint, *if* the capacity of the network is optimally adjusted to the level of demand *and* there are constant returns to scale[2], congestion costs will signal the need to expand capacity *when* the latter costs exceed the cost of capital. Note that congestion costs are positive (not zero) when capacity is optimal, a point usually neglected when the costs of congestion are the subject of popular debate. Consequently, starting with an assumption of a constant cost network optimally adjusted, the EC proposal for charges based on SRMC, where the latter include the costs of congestion, will allow for an adequate financial return.

This conclusion does not hold of course when there are scale economies but this only causes a problem in this context if there are economies (as opposed to diseconomies) of scale. My view is that the UK rail industry (and probably rail industries generally) is a constant or even an increasing cost industry; it is expensive to add an increment of capacity to what is a fairly dense network and US academic studies tend to support this conclusion. This is not to deny the existence of economies of density or utilisation, but this characteristic bears upon ex-post prices; the skill for the commercial railway will be to learn to add small increments to capacity so that capacity is generally, and constantly, adjusted to demand. If capacity is not so adjusted, this is only a problem in the current context when there is too much capacity. But I am of the view that the dominant feature of the UK network is not excess capacity but the contrary, congestion is endemic; the majority of train-miles operate over congested track (in London and the south east, around the major conurbations and along the intercity network)[3].

In its contribution to the preparatory work for privatisation, BR argued that widespread capacity constraints were one of the most important obstacles to the introduction of on-rail competition in train services. BR put forward a telling example to illustrate the point choosing a relatively simple part of the network in the early hours of the morning:

> ... the critical aspect is that even at 3 and 4 o'clock in the morning, we have a whole series of trains which have to wait. One freight train has to wait over an hour off the main line, in order to allow a "flight" of sleeping car trains to pass. In fact, at this time all three passing loops are occupied to enable the passenger trains to go through. So even very early in the morning, there are capacity constraints on this simple piece of the network. The question of running an additional train involves examining a series of inter-dependencies to see if we can re-jig the operation to allow it through without upsetting existing services. (John Welsby, 1991)

The circumstance set out in the quote, in spite of its idiosyncrasy meets the requirement in Article 8 of the EC proposal, that a charge reflecting the scarcity of track capacity 'shall only be levied on identifiable segments of infrastructure which are subject to capacity constraints'. In this example, therefore, and I would argue that for much of the network, the application of the EC proposal for SRMC pricing would produce a surplus of revenues in excess of wear and tear and related costs.

Whether the aggregation of the congestion charges and scarcity rents over the network as a whole would be sufficient to provide an adequate return for Railtrack's shareholders and provide sufficient funds for enhancing the network is, to a large extent, an empirical question. But my intuition is that an appropriate charging mechanism based on economic concepts of congestion will do much to produce an adequate cash flow[4]. There will be parts of the network which will prove to be uneconomic when measured in cash flow terms using SRMC as a basis for charges, (including Chris Bolt's testing example of the freight only line shared between two users) but such parts will have a call on the scarcity rents generated by the 'long term bottlenecks', those sections of track where demand exceeds capacity but are expensive to expand so that the costs of expansion exceed the social benefit.

Bear in mind also that there are circumstances when a deviation from strict SRMC pricing is permissible under the EC proposals: access charges may be based on the long-run additional costs arising from the particular investments, albeit in exceptional circumstances and subject to certain criteria; and SRMC prices can be adjusted in various ways in order to achieve a higher level of cost recovery.

In the round, therefore, I am reasonably optimistic that the EC proposals will meet the benchmark set out initially. Certainly what is offered remains a substantial improvement on what we have in place at the present time.

Road Pricing and Road Finance

David Newbery's paper first discusses public finance issues relating to roads and in this context he constructs a set of accounts for the UK road network for 1996/97. The essential point that emerges is that road taxes (revenues) after allowing for general consumption taxes (VAT), exceed costs by a considerable margin. There have been a number of studies in the past which have constructed similar accounts and these have produced varying answers in view of the constantly changing road tax and expenditure regime[5]. However, in this instance I do not disagree with the conclusion although there are some aspects of the calculation which might lead to costs being underestimated. For example, not all the costs of road accidents are internalised in private motoring costs, a residual falls to the public purse; and current maintenance expenditure probably underestimates the true costs of maintaining the network in a steady-state condition. Also the imputed costs for environmental externalities of 1p/PCU km is possibly on the low side. Thus the imbalances between revenue and costs may not be as great as David Newbery suggests but I am willing to accept that an excess of revenues does exist.

This excess is taken by Newbery to indicate that there should be more spending on increasing the supply of road infrastructure (although this would not follow if the road supply industry was an increasing cost industry in which case we would expect a surplus of revenue over costs). The imbalance between spending and taxes is considered to be particularly marked in the case of inter-urban roads. In this context Newbery dismisses the case for tolling commenting that: '...an efficient road transport policy would rarely need to resort to road pricing on inter-urban roads and motorways', a view based on the observation that '...congested inter-urban roads are sufficiently cheap to expand that current road taxes more than cover the costs of expansion per [vehicle kilometre] often by a factor of two'.

However, I would question this viewpoint. In the case of the inter-urban motorway network, capacity has been added at the margin most recently by widening existing motorways as part of a large scale programme set in train by the previous Government. But these widening schemes are extremely expensive on a lane/km basis compared with a base-line lane/km cost for new motorway construction. Unpublished figures submitted by the then Department of Transport's National Roads Policy Directorate to the Transport Select Committee in 1994, indicated that the cost of widening motorways was almost three times the cost per lane of a new three lane motorway. Moreover the cost of widening was considered highly variable and that if rebuilding of junctions was involved, then the cost could be around half as much again.

This leads me to four conclusions. First, it is likely that the roads industry is, in general, an increasing cost industry. Secondly, it is possible that the price for road use is frequently less than the marginal costs of supply[6], in which case both the reining in the inter-urban roads programme by the Government and a policy of tolling motorways might be justifiable on grounds of economic efficiency. Thirdly, the European differences that David Newbery draws attention to, might be partly explained by differences in the marginal costs of supply in the different countries. For example, we might expect the lower population density and larger land mass of Spain and France to be a factor in the lower costs of road supply in these two countries (although this factor does not explain the relatively low costs in Belgium or Italy). And, fourthly, with evidence from the motorway widening programme in mind the indications are that the marginal costs of supply varies significantly between locations and this might be a further justification for inter-urban road pricing.

Finally, I am conscious that I have not thus far commented directly on urban road pricing which forms a substantial part of Newbery's paper. I have little to add here, both to his analysis and to his policy conclusions. The research that is being undertaken appears meticulous in its regard to complex technical relationships between traffic densities and speed in urban networks. The only points I wish to make is that the consideration of the peak/off-peak issue might also benefit from consideration of the firm peak/shifting peak concept (particularly in view of our knowledge about peak spreading) and that for testing road pricing systems perhaps serious consideration could be given to using the City of London.

The City already has, as a result of the security cordon, a restricted number of access points; through traffic has been diverted around the area[7] and much of the terminating traffic has free parking in privately controlled parking spaces. Both politics and economics might suggest that the introduction of cordon pricing in the City and elsewhere, could be accompanied by the removal of parking meters (but not designated on-street parking spaces). If it is possible to control traffic volumes by prices at the cordon there is less need to ration on-street parking by price and all parking is then treated equally. Removing metered parking might also reduce opposition to the idea of efficiently pricing urban roads.

References

Jennings, A. (1979) Determining a Global Sum for the Taxation of Road Users *Journal of Transport Economics & Policy*, 13,1.

Starkie, D.N.M. (1979) Allocation of Investment to Inter-Urban Road and Rail *Regional Studies,* 13,4.

Welsby, J. (1991) Rail Privatisation: British Rail Imperatives *Proceedings of the Ninth Annual Conference*, Major Projects Association, September 1991.

Notes

1   The provisions of the 1993 Railway Act apply in England, Scotland and Wales but not to Northern Ireland.

2   So that SRMC equals LRMC (or its Long Run Incremental Cost approximation).

3   This is more apparent if congestion is defined from an economic perspective rather than adopting the traditional approach which is to refer to the notional capacity of the network. Congestion is manifest when the optimal timing of a particular service (from the viewpoint of its operator) has to be adjusted to fit in with the timing preferences of other operators. In these circumstances there is an opportunity cost caused by the "crowding out" of one service by another.

4   The real challenge of the proposals will be in establishing a satisfactory methodology for measuring congestion; the proposed directive does not specify how these charges should be calculated.

5   See for example Jennings (1979) for a review of Government calculations and Starkie (1979) for an attempt to calculate accounts for the inter-urban network.

6   If the average cost of widening per lane/km is three times that for an average new-build three lane motorway this would suggest a cost per vehicle/km of about 7p using figures in Newbery (1994) as a basis. This might be compared with an average 1998 revenue per vehicle/km of about 5p.

7   A major criticism of cordon pricing is that it tends to transfer traffic to areas outside the cordon. With the City of London security cordon this transfer has already occurred.

6   How Effective is Competition? The Case of Rail Services in Britain

JOHN PRESTON[1]

Introduction

Britain's railways were reformed as a result of the 1993 Railways Act and related legislation. The main provisions of this Act are documented elsewhere (see, for example, Root and Preston, 1998). Although much of the Act was related to reform of organisational structures and ownership, there were also important provisions for increased competition. Competition off the tracks (or for the tracks) was introduced for the passenger business through the medium of franchising. Competition on the tracks for domestic passenger services was limited initially, although these limitations are due to be relaxed in September 1999 and possibly lifted completely in 2002. Competition for domestic rail freight services was permitted by the 1993 Act whilst EC directive 91/440 made provision for competition for international passenger and freight services. The 1993 Act also permitted organisational reform of the railway supply industry in order to make it more competitive. This paper will focus on competition in the passenger rail sector but will consider other aspects briefly in the concluding sections.

First, the provisions for the first round of franchising and for open access are reviewed. Secondly, a research methodology used to assess on-track and off-track competition is described. Thirdly, the results from our work on off-track competition are presented, whilst, fourthly, our results from our work on on-track competition are presented. Fifthly, a brief review of the overall impact of competition and the other reforms is provided. Lastly, some tentative conclusions are drawn.

Background

A near unique aspect to British Rail's privatisation was the transfer of businesses that had little chance of making a profit to the private sector. In

1993/4 British Rail's passenger business as a whole only covered 72% of its costs, based on the accounting conventions then in use (BRB, 1994). Moreover, subsequent changes to the charging regime for infrastructure and rolling stock led to the belief that only one passenger business (Gatwick Express) would be profitable at the outset of the privatisation process (Dodgson, 1994). The solution to this problem was one which was first proposed in general terms by Chadwick (1859) and revived by Demsetz (1968), namely to hold an auction for natural monopolies in services. However, for passenger rail operations in Britain, such auctions would involve bids with negative prices (i.e. subsidy is required), and has become known as franchising.

The passenger business of British Rail was horizontally separated into twenty-five geographically based train operating companies. These businesses were then privatised via a franchising regime administered by the Office of Passenger Rail Franchising (OPRAF). The announcement of the first franchises to pass into the private sector was made in December 1995, whilst the announcement of the completion of the franchising process was made in February 1997.

Once certain specific quality thresholds were met, franchises were awarded to those companies that asked for lowest amount of subsidy. The results of this process are presented in Table 1. The main feature of this Table is that subsidy in the first year is forecast to be around £2.1 billion. This compares to the £545 million revenue subsidy the industry received in 1993/4. This refers to Public Service Obligations (PSO) and Section 20 payments from Passenger Transport Executives only. If capital grants, restructuring grants and level crossing grants are included, the subsidy total for 1993/4 increase to £1086 million. By the end of the franchise periods (which range from 2003 to 2012) the subsidy will be reduced to around £530 million (i.e. very close to the 1993/4 level for PSO and Section 20 payments).

Although a relatively small number of organisations were involved in the bidding, it had the appearance of being very competitive, with typically five serious bids per contract. The successful bidders are predominantly bus operators, (fifteen out of twenty-five franchises) with a small number of management buyouts (four franchises). Virgin, a French conglomerate (Companie Generale des Eaux), Sea Containers and a consultancy-led company (GB Rail) were the other successful purchasers. The dominance of bus companies has raised concern about the lack of competition where the new railway franchise operator is also the major bus operator in particular districts.   However, it should also be noted that this situation has

Table 1 The franchise awards (current prices)

Franchise	Franchisee	Franchise Length (years)	Subsidy Year One (£m)	Subsidy Final Year (£m)
Great Western	MEBO with 3I & First Bus	10[†]	61.9	28.3
South West Trains	Stagecoach Holdings Plc	7	63.3	35.7
East Coast Main Line	Sea Containers	7	67.3	0
Midland Mainline	National Express	10[†]	17.6	-10.2
Gatwick Express	National Express	15[†]	-4.1	-23.1
LTS Rail	Prism	15[†]	31.1	11.7
South Central	Connex	7	92.8	35.9
Chiltern Railways	MBO with 3I and John Laing	7	17.4	3.3
South East Trains	Connex	15	136.1	-1.3
South Wales & West	Prism	7yrs 6mths	84.6	39.2
Cardiff Railways	Prism	7yrs 6mths	22.5	13.6
Thames Trains	Go-Ahead with MBO	7yrs 6mths	43.7	0
Island Railways	Stagecoach	5	2.3	1.8
North Western Regional Railways	G&W Holdings	7yrs 1mth	192.9	125.5
North East North London Railways	MTL	7yrs 1mth	231.1	145.6
	National Express	7yrs 6mths	55.0	15.8
Thameslink	GOVIA	7yrs 1mth	18.5	-28.4
West Coast Main Line	Virgin	15	94.4	-220.3
ScotRail	National Express	7	297.1	202.5
Central	National Express	7yrs 1mth	204.4	132.6
CrossCountry Trains	Virgin	15	130.0	-10.3
Anglia	GB Railways	7yrs 3mths	41.0	6.3
Great Eastern	FirstBus	7yrs 3mths	41.3	-9.5
WAGN	Prism	7yrs 3mths	72.6	-25.5
Merseyrail	MTL	7yrs 2mths	87.6	60.8
TOTAL			2102.4	530.0

Notes:

[a] Negative Subsidies indicate payment of a premium.

[b] [†] Conditional on delivery of franchise plan commitments on rolling stock investment.

Source: OPRAF Annual Reports and Accounts 1996-97

opened up some opportunities for single companies to run interconnecting train and bus services where previously such routes were not provided (Willich, 1996). An example is in the Oxford area where the Go-Ahead Group operates Thames Train services and one of the main bus companies in the area.

The franchising process has resulted in a major concentration of ownership amongst a few operators. By revenue, the largest operations were initially Connex, Virgin, National Express and Stagecoach, accounting for 54% of all revenue. However, there has already been further agglomeration. Virgin and Stagecoach have effectively formed an alliance, with Stagecoach holding 49% of Virgin Train's shares, whilst First Group (formerly First Bus) have taken over Great Western Holdings to become a member of the big four, which now control 70% of the passenger rail market. Similar agglomeration (albeit by the route of direct sales) has occurred in the rail freight market, where English, Welsh and Scottish, a subsidiary of Wisconsin Central, acquired five out of the six major companies sold, although there are also some own account rail freight operators. With regard to on-track competition, the Rail Regulator has so far limited competition to routes accounting for less than 0.2% of a franchisee's revenue on flows registered with ORR or on which no through service is operated. The current proposal is that, following a new flow registration process, this should rise to 20% of nominated flows (which must include already competed flows) in September 1999 (ORR, 1998). A further review of policy with the possibility of continued incremental change would take place in 2002 (ORR, 1994). Actual on-track competition in the privatised market has therefore been restricted to areas where franchises overlap (e.g. Connex South Central, Gatwick Express and Thameslink on the Gatwick-London route) and areas where parallel routes exist (e.g. London-Birmingham).

Methodology

The starting point of our research centred on extensive data collection. This exercise involved undertaking 428 stated preference interviews and collecting 1,532 revealed preference questionnaires. The aim was to focus on two forms of competition: competition for the market (off-track competition) and competition in the market (on-track competition).

Off-Track Competition

Between May and November 1995, we conducted in-depth interviews with 38 potential franchisees; 20 of whom were directors of British Rail Train Operating Units, 7 from large bus companies and 11 from other rail-based institutions (e.g. OPRAF, Railway Industry Association, Railtrack). Although somewhat biased towards former British Rail directors, on analysis our sample included decision-makers from 8 of the 13 successful bidders. The purpose of the interviews was to obtain bidding preferences for alternative franchise specifications in a hypothetical bidding experiment as well as discussing more generally issues arising from the privatisation of British Rail. The overall objective was to identify what would make for an attractive franchise from a franchisee's perspective.

From earlier review work, we identified four attributes of a franchise that were worthy of detailed quantitative analysis in a Stated Preference (SP) bidding experiment. These were:

- subsidy requirement;
- contract length;
- exclusivity (with and without open access competition); and
- degree of regulatory control.

The design was customised in that respondents could choose from experiments for five different franchises: South West Trains, Chiltern, Inter City East Coast (ICEC), Inter City West Coast (ICWC) and ScotRail. This, in effect, meant that a fifth attribute, that of contract type/size, could be examined. This work is described in detail by Preston and Whelan, 1995, 1996, and Preston *et al.*, 1997.

On-Track Competition

The second tranche of survey work was aimed at the development of a model to be used in the analysis of on-track competition. To facilitate an assessment of the demand implications of open access competition we needed to be able to forecast how rail travellers will respond to changes in train timetables, fares, ticket availability, journey time and interchange requirement. This information allowed us to assign any given person with given desired outward and return leg departure times to particular services and ticket type.

Information on passengers' ideal departure times for both outward and return legs of their journey and information on passengers' sensitivity to a host of rail attributes were gathered via an extensive data collection exercise involving a self-completion questionnaire and two computerised stated preference (SP) interviews. The self-completion questionnaire survey was conducted on board all trains operating on the study route on Tuesday 27 May 1995. 'Top-up' surveys for missed trains were conducted on Tuesday 5 September 1995. A total of 1531 responses were obtained. This survey provided the data for our revealed preference (RP) models and provided the basis of our forecasting procedure which is based on sample enumeration.

The first of the two SP experiments was aimed at assessing passengers' choice of ticket type. It involved two sections. Firstly, introductory questions sought information on preferred departure times for both legs of the journey, ticket type and other socio-economic characteristics. This information was then used in the design of a stated preference exercise presented in the second section of the questionnaire. This exercise asked respondents to choose between open, saver and APEX tickets, each offering different fares, advanced purchase requirement and time restrictions on usage, and not travelling by train.

The second of the two SP experiments was aimed at assessing passengers' choice of class and mode of travel. Once again the experiment involved two sections. The first section collected background information on journey and individual characteristics as above, but also on class of travel and access times to stations. The second section of the questionnaire presented a stated preference exercise offering a choice between train first class, train standard class, car and either air for business travellers or coach for non-business travellers. The overall approach thus involved SP data on choice of mode, SP data on ticket type, and RP data on choice of ticket type and mode.

Results - Off-Track Competition

Our assessment of off-track competition draws on data collected during the in-depth interview process conducted with potential franchisees and described above. In total, the hypothetical bidding game yielded a data set of 511 preferences and 1022 bids from 33 respondents. This data was analysed using a multinomial logit model in order to establish managers' preferences with respect to contract size and length, exclusivity, and the degree of regulatory control. The results from the model are shown in Table 2.

The model has a respectable overall fit and has intuitively correctly signed coefficients that, with one exception, are significant at the usual 5% level. Parameter estimates show that there is a preference for longer franchises. We estimate that extending franchises by five years (from 7 years to 12 years) would reduce subsidy requirements for an average franchise by around £3.8 million per annum. There was a strong preference for franchises to be exclusive, as this would reduce the subsidy required for

Table 2 Results of the franchising SP experiment

VARIABLE	Coefficients and associated t-statistics (in brackets)				
	ICEC	ICWC	SCOTRAIL	CHILTERN	SOUTH WEST
Franchise Dummy	-3.181 (3.1)	-6.295 (3.6)	-35.78 (8.6)	-11.68 (8.2)	-11.68 (8.2)
Subsidy	0.1118 (4.1)	0.1118 (4.1)	0.1931 (8.8)	0.3568 (9.1)	0.1931 (8.8)
Franchise Length	0.0776 (2.0)	0.1750 (2.3)	0.01718 (0.4)	0.3083 (5.4)	0.1084 (3.1)
Exclusivity	0.6220 (2.3)	1.222 (6.0)	1.222 (6.0)	1.222 (6.0)	1.222 (6.0)
Regulation	-0.4922 (2.6)	-1.282 (4.1)	-1.282 (4.1)	-2.495 (5.4)	-0.4922 (2.6)
% of responses	30	7	18	17	28
No. of Observations	1022				
Rho Squared	0.1690				

a typical franchise by around £6.5 million per annum. There was some evidence to suggest that open access competition (i.e. non-exclusivity) is most expected on Inter City routes. The proposed regulatory regime was seen by our interviewees as being excessive. It is estimated that a more liberal regime would lead to reductions in subsidy for a typical franchise of around £6.4 million per annum. Overall, our analysis suggests that a move to longer (12 year), exclusive and loosely regulated franchises could lead to an annual subsidy reduction of up to £415 million compared to a regime based on non-exclusive, tightly regulated, seven year franchises (a decrease in the forecast total subsidy bill of some 25%). In the event, seven of the 25 franchises have been awarded for 10 years or more, whilst some form of exclusivity has been guaranteed until 2002.

Analysis of data from the in-depth interviews told us that competition for franchises was expected to be relatively intense with 3 to 5 bids per franchise, one of which would be a Management Buy-Out (MBO) - this has indeed been the case. A period of consolidation was expected with the industry re-agglomerating into around four groups. We have seen that this has already happened. This may have implications for off-track competition in the future. Our analyses suggests that if there are only one or two bids per franchise subsidy requirements are likely to increase. Our results suggest that bidders prefer relatively self-contained and/or recently upgraded routes, and that routes which require substantial investment are particularly problematic. Routes of this type turned out to be the last ones to be franchised. Our interviews also indicated that those from outside the industry were more bullish about the prospects for cost reductions and revenue increases than insiders. This has been validated by subsequent events in that only three of the 25 franchises have been won by MBOs.

Winning bid forecasts based on up-to-date financial information were estimated and validated with data on actual bids. Initially forecasts were made for the five franchises outlined in the experiment but subsequent forecasts have been made for all franchises. For franchises not covered in the experiment, forecasts were made by applying the parameter estimates of a closely resembling 'experiment franchise' and adjusting the franchise specific constant to take account of pre-privatisation base subsidy requirement. Table 3 shows the results of this exercise.

It can be seen that franchises let at the outset of the franchising program were generally awarded for less than their forecast 'market' value, whereas franchises awarded towards the end of the process were awarded for substantially higher than their forecast market value. Anecdotal evidence explaining this phenomenon suggests that the degree of risk associated with making a bid during the initial franchising tranches was high due to high levels of uncertainty surrounding the process.   Potential franchisees therefore needed to be compensated for bearing this risk.   As the franchising process advanced, however, players began to understand the system thereby reducing uncertainty. This reduced risk, coupled with an increase in the likelihood that if potential franchisees were unsuccessful they may have to wait seven years to bid for another, lead to bids becoming progressively more optimistic.

It can be seen that the order of letting is almost significant at the 5% level and, all other things equal, the tenth group of franchises let require £18 million less subsidy per annum than the first group. This finding is similar to that detected by Harris (1997) who noted that the potential for growth, the award of longer franchises to those bidders offering new trains and the order of letting provide significant explanatory power in determining subsidy requirements.

On the basis of this analysis the winner with the biggest task appears to be Virgin. We estimate that the subsidy they will receive from Government falls short of what might be required by as much as £130 million perannum. The biggest gainer is Stagecoach who, we estimate, are receiving around £27 million per annum more than is required. The alliance with Stagecoach

Table 3   Forecasts of winning bids (£m)

FRANCHISE	Forecast Winning Bid	Actual Average	Actual - Forecast	Order of Letting
Great Western	23.37	45.10	21.73	1
South West Trains	28.85	49.50	20.65	1
Gatwick Express	-21.10	-13.60	7.40	2
East Coast Main Line	26.71	33.65	6.94	2
Midland Mainline	-10.79	3.70	14.49	2
LTS Rail	15.87	21.40	5.53	3
South Central	75.35	64.35	-11.0	3
Chiltern Railways	23.97	10.35	-13.62	4
Cardiff Railways	15.60	18.05	2.45	5
Island Railways	-3.39	2.05	5.44	5
South East Trains	101.58	67.40	-34.18	5
South Wales & West	81.34	61.90	-19.44	5
Thames Trains	35.06	21.85	-13.21	5
Anglia	29.90	23.65	-6.25	6
Cross Country	102.14	59.85	-42.29	6
Great Eastern	24.33	15.90	-8.43	6
W.Anglia & Gt.Northern	49.85	23.55	-26.30	6
Merseyrail	73.10	74.20	1.1	7
Central	193.59	168.50	-25.09	8
North London Railways	37.45	35.40	-2.05	8
Regional Railways NEast	221.45	188.35	-33.1	8
Regional Railways NWest	198.62	159.20	-39.42	8
Thameslink	0.52	-4.95	-5.24	8
West Coast Main Line	25.16	-62.95	-88.11	9
ScotRail	313.16	249.80	-63.36	10
TOTAL	1661.69	1316.20	-345.49	

The hypothesis that subsidy levels decrease with the order of letting was tested by regressing actual average subsidy required over the length of the franchise against forecast subsidy requirement and order of letting. The results of the regression are shown in the equation below (t-statistics shown in brackets).

Actual Average = 0.889 Forecast Average - 1.972 Order of Letting   (Adjusted R^2 0.9374)
                   (14.881)   (1.859)

seems a sensible strategy for Virgin. It should be noted that an important element missing in our forecasting model is an assessment of revenue growth as a result of initiatives such as new rolling stock. This may provide another explanation as to why our forecast bids for Virgin are so much higher in terms of subsidy than the out-turn.

Results - On-Track Competition

On the demand side, the three different data sets described in the methodology section were analysed so as to build a disaggregate demand model examining the choice of ticket type, class of travel and mode of travel. On the supply side, an accountancy cost model detailing both capital and operating costs was specified on the basis of earlier work at Leeds by Galvez (1989) and Worsey (1994). The two models coupled together provide a methodological base for the assessment of on-track competition. We have calibrated a model with both RP and SP data, we have linked our model with the overall demand for rail travel, we have considered APEX fares and we have considered both legs of the journey whereas previous models just looked at one leg.

The template for the operations model is an actual intercity rail line in Great Britain. For ease of modelling we have made some simplifying assumptions. The route is treated as a self-contained unit incorporating eight stations with no infrastructure capacity constraints. The resultant eight by eight demand matrix is based on actual point to point demand information obtained from ticket sales data. Appropriate adjustment of these figures was taken using survey data to account for passengers travelling on the route for only part of their journey. To simplify modelling, individuals are assumed to make their travel decisions at three linked stages (Figure 1).

The model combines both demand and cost elements under a user-friendly 'front end'. The analyst is free to specify any combination of services from up to three operators, each having different fares and ticketing restrictions. Model output includes demand and revenue for each service, service costs, operator profit, consumer surplus and economic welfare estimates. Examples of the model's output are given by Tables 4 to 10. Further details are given by Whelan et al, 1997A, 1997B and Preston et al., 1998.

Taking the incumbent's May 1995 service pattern and fare structure as the base situation we looked at four possible scenarios for duopolistic on-track competition.

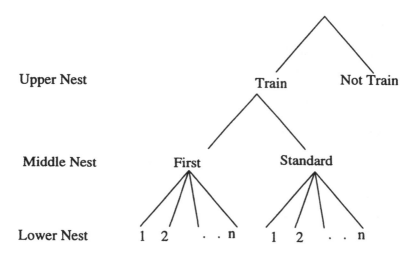

Figure 1   Schematic representation of the demand models

Cream Skimming

The first competitive situation to be assessed was that of cream skimming. Here we assume that the entrant is able to 'cherry pick' peak services without the obligation to operate possible loss-making services in the off-peak.   Under current proposals, this scenario may be possible from September 1999 onwards. To begin with it was assumed that the entrant augments the incumbent's timetable providing two additional services in the morning and evening peak periods in both directions of travel. Next it was assumed that infrastructure capacity constraints exist and the entrant simply replaces two of the incumbent's services in the morning and evening peak periods. Both scenarios were specified with and without ticket inter-availability (i.e. passengers with inter-available tickets are able to use different operator services on different legs of their journey) and both with and without fare discounting on the part of the entrant. The results are shown in Table 4. It should be noted that in the base situation, this rail service was estimated to be highly profitable, with revenue of £137,481 per day and costs of £94,735, resulting in daily profits of £42,746. It should also be noted that we are only modelling the core part of a franchise on a weekday. In practice, this profit may be used to cross-subsidise weekend travel and travel on other peripheral services provided by the franchisee.

It can be seen that cream skimming entry is likely to be profitable (especially where the entrant offers discounted fares) and results in a possibly sustainable environment in which both operators are able to make a profit. Where additional services are provided, business and non-business passengers are shown to be better off as indicated by higher levels of consumer surplus. However, passengers are shown to be worse off when the incumbent is forced to surrender train paths to the entrant and there are no offsetting fare reductions. This is because stopping services are replaced by express services with the consequence that some stations are served by fewer trains. There appears to be strong commercial incentives for the entrant to discount fares and for the incumbent to not accept tickets issued by other operators. Where the entrant discounts fares, it is likely that the incumbent will also discount. The market shows an overall welfare loss in all scenarios, with one exception (model run 11) in which existing services are operated but with 20% lower fares. These falls in welfare are due to heavy profit losses on behalf of the incumbent and indicates that the market is inelastic with respect to service frequency, but less so with respect to fares. At base fares and service levels our model suggests a fare elasticity of -0.5 for business and -1.0 for leisure which is in line with industry standards (Rail Operational Research, 1997).

An important issue highlighted by Table 4 is the distribution of benefits. In a number of model runs (e.g. 5 and 12), there are large transfers of benefits from producers to users, whilst the producers remain profitable.

This may be seen as a politically desirable result. A related issue is the extent to which competition reduces the scope for temporal and spatial cross-subsidy and the resultant financial and welfare impacts.

Head-on Competition

The second competitive situation to be assessed was that of head-on competition. In this scenario it is assumed that the entrant matches the service frequency of the incumbent. Each operator is assumed to operate alternate trains throughout the day. The results of this simulation exercise are shown in Table 5.

Table 4 Sample cream skimming simulation results (£ per day)

Model Run	Fares/ Entrant Service Pattern	Inter- Avail- ability of Tickets	Incumbent Profit	Entrant Profit	Consumer Surplus Change (Business)	Consumer Surplus Change (Non- business)	Welfare Change
1	A/1	Yes	30,815	1,267	1,528	82	-9,051
2	A/1	No	31,962	-847	891	82	-10,657
3	B/1	Yes	12,419	16,670	4,686	791	-8,178
4	B/1	No	17,799	10,379	3,510	512	-10,544
5	C/1	Yes	23,545	528	12,741	4,548	-1,383
6	C/1	No	25,017	-2,135	12,055	4,483	-3,326
7	A/2	Yes	29,591	11,381	-3,578	-1,046	-6,397
8	A/2	No	29,553	9,183	-4,603	-1,153	-9,765
9	B/2	Yes	20,050	18,888	446	-210	-3,570
10	B/2	No	22,158	14,700	-845	-507	-7,239
11	C/2	Yes	23,241	10,259	7,592	3,380	-1,727
12	C/2	No	23,240	7,999	6,466	3,230	-1,810

Notes:

1   Entrant provides two additional express services in the morning and evening peak periods in both directions of travel.

2   System is at capacity, the entrant replaces two of the incumbent's services in the morning and evening peak periods in both directions of travel with express services.

A   Entrant price matches incumbent's base fare levels

B   Entrant discounts fares by 20%

C   Both operators discount fares by 20%.

It is clear from Table 5 that in this competitive environment the entrant relies on fare discounting to make a profit. The incumbent is likely to suffer a drastic reduction in profits and will be forced to respond to remain profitable. Although this form of competition is welfare negative due to profit loss on behalf of the incumbent, the increase in service frequency and reduced fares are beneficial to passengers. This scenario is unlikely to be sustainable and may result in a price war.

Table 5 Sample head-on competition simulation results (£ per day)

Model Run	Fare Difference (Entrant)	Inter- avail- ability of Tickets	Incumbent Profit	Entrant Profit	Consumer Surplus Change (Business)	Consumer Surplus Change (Non- business)	Welfare Change
13	0	Yes	5,826	-11,472	8,436	3,747	-36,208
14	0	No	5,881	-13,906	6,113	2,687	-41,970
15	-10%	Yes	-12,823	3,821	10,937	5,033	-35,778
16	-10%	No	-9,077	-1,803	8,410	3,960	-41,256
17	-20%	Yes	-30,271	14,982	14,308	6,726	-37,001
18	-20%	No	-25,470	8,898	11,420	5,641	-42,255

Note: Entrant matches service frequency of incumbent

Price War

The third competitive situation to be tested involves a price war. It is a natural development to head-on competition and we assume the entrant to be the price leader.

It can be seen from Table 6 that the entrant can only make a profit where fares are discounted at a greater rate than the incumbent and tickets are inter-available. To retaliate, the incumbent need only to price match or withdraw inter-availability. This scenario appears to be unsustainable and it is likely that the entrant would be forced out of business. As in the analysis of head-on competition the market experiences a welfare loss in all competitive scenarios even though consumers gain considerably. Again equity emerges as an important issue.

Product Differentiation

We examined quality competition by examining the prospects for a slow but cheap service running on a parallel route. This is competition between the tracks rather than on the tracks. In these circumstances, the entrant could capture a significant market niche, namely early morning non-business travellers. With fares at 50% of those on the incumbent's services, the parallel entrant could capture 25% of the rail market. We were not able, however, to undertake a full welfare assessment of this option as this would require data on the demand and cost characteristics of the entrant's operations which were not available to us. It is interesting to note that this form of competition has already arisen on some routes in the form of discounts for group travel, loyalty bonuses and improved marketing initiatives through telesales and the internet.

Alternative Track Charging Systems

The model runs described above were based on a fully allocated costing system for rail infrastructure, which results in a decline in access charges for the incumbent following entry. An alternative set of model runs were undertaken in which the entrant was only charged for the marginal cost of the infrastructure. The results are shown by Table 7. Although this charging system would make entry more likely, the incumbent would make losses in excess of the entrant's profits in all scenarios examined whilst the welfare change remained negative. Table 8 shows that, where both the entrant and the incumbent are charged marginal costs for the infrastructure, head-on competition is feasible but the infrastructure authority is now loss making.

Table 6 Sample 'price war' simulation results (£ per day)

Model Run	Fare Incumbent Entrant	Inter-avail-ability of Tickets	Incumbent Profit	Entrant Profit	Consumer Surplus Change (Business)	Consumer Surplus Change (Non-business)	Welfare Change
19	0/0	Yes	5,826	-11,472	8,436	3,747	-36,208
20	0/0	No	5,881	-13,906	6,113	2,687	-41,970
21	0/-10%	Yes	-12,823	3,821	10,937	5,033	-35,778
22	0/-10%	No	-9,077	-1,803	8,410	3,960	-41,256
23	-10%/-10%	Yes	4,254	-13,677	13,954	5,890	-32,325
24	-10%/-10%	No	4,083	-15,813	11,645	4,835	-37,995
25	-10%/-20%	Yes	-14,547	1,347	16,343	7,228	-32,375
26	-10%/-20%	No	-11,010	-3,961	13,832	6,155	-37,730
27	-20%/-20%	Yes	1,724	-15,943	19,417	8,104	-29,444
28	-20%/-20%	No	1,467	-17,902	17,117	7,054	-35,010

Another possibility would be for the timetable to be split into two packages and the two operators bid for track access, although this was envisaged as a pre-cursor to evolving competition (Starkie, 1993). The possible impacts of such a policy are investigated for the base timetable in Table 9.

It can be seen that our model assumes that under these circumstances costs are divisible and, given perfect information and perfect competition for the bids, the combined bids for the halves of the combined timetable should be the same as the bid for the single timetable, provided tickets are inter-available. However, it is noticeable that one half of the timetable is worth more than the other half (operator A's surplus is 26% higher than operator B's). In cases of imperfect information, this is more likely to be known by the incumbent than the entrant. The naïve entrant might bid up to an amount equivalent to £40,036 per day (i.e. around half the operating surplus for the integrated timetable), winning operator B's services and making a loss of £4,646 per day. An incumbent with perfect information could shave the entrant's bid for operator A's services and make a profit of £4,645 per day. When the ticketing is no longer inter-available, total operating profit falls suggesting one combined bid would usually win over two separate bids. This analysis suggests that for the Starkie model to work there needs to be perfect information, divisible costs that do not exhibit returns to scale and regulation to ensure integrated ticketing. It also assumes that there is no scope for product differentiation in terms of fares and comfort (e.g. in terms of carriage layout).

Table 7   Sample head-on competition simulation results (£ per day) assuming marginal cost infrastructure pricing for the entrant only

Model Run	Fare Difference (Entrant)	Inter-availability of Tickets	Incumbent Profit	Entrant Profit	Consumer Surplus Change (Business)	Consumer Surplus Change (Non-business)	Welfare Change
29	0	Yes	-9,417	3,951	8,436	3,747	-36,208
30	0	No	-8,842	817	6,113	2,687	-41,970
31	-10%	Yes	-34,049	25,047	10,937	5,033	-35,778
32	-10%	No	-28,637	17,757	8,410	3,960	-41,256
33	-20%	Yes	-45,253	41,839	14,308	6,726	-37,001
34	-20%	No	-50,329	33,757	11,420	5,641	-42,255

Note: Entrant matches service frequency of incumbent

Table 8   Sample head-on competition simulation results (£ per day) assuming marginal cost infrastructure pricing for the incumbent and the entrant

Model Run	Fare Difference (Entrant)	Inter-availability of Tickets	Incumbent Profit	Entrant Profit	Consumer Surplus Change (Business)	Consumer Surplus Change (Non-business)	Welfare Change
35	0	Yes	27,911	3,951	8,436	3,747	-36,208
36	0	No	28,486	817	6,113	2,687	-41,970
37	-10%	Yes	3,279	25,047	10,937	5,033	-35,778
38	-10%	No	8,691	17,757	8,410	3,960	-41,256
39	-20%	Yes	-7,925	41,839	14,308	6,726	-37,001
40	-20%	No	-13,001	33,757	11,420	5,641	-42,255

Note: Infrastructure authority loses £37,328 per day
Entrant matches service frequency of incumbent

In a second case study our methodological framework for assessing open access competition was applied to a London and South East route. This study employed similar demand, cost and evaluation models but made appropriate adjustment for differences in attribute values and market elasticities. Unlike the inter-urban case study where competitive scenarios were selected a priori, a broad range of pricing, output and quality strategy combinations, totalling 1,296 scenarios, were simulated. The presentation of this amount of data however would be cumbersome and difficult to interpret. Therefore, the outcome of all tested scenarios were analysed via a series of dummy variable regression runs. In the first set of models, operator profitability was taken as the dependent variable and the strategies of the entrant and the incumbent taken as independent (dummy) variables.

Table 9   Impact of bidding for timetable packages

Scenario		Revenue (£ per day)	Operating surplus (£ per day)
Integrated timetable and tickets	Operator A	137,481	80,073
	Total	137,481	80,073
Separated timetable, integrated ticketing	Operator A	75,380	44,682
	Operator B	62,100	35,390
	Total	137,480	80,072
Separated timetable and ticketing	Operator A	74,033	43,338
	Operator B	58,801	32,089
	Total	132,834	75,427

The regression models therefore form general profitability relationships for each operator where the coefficients indicate what strategy makes sense for each operator to undertake given the range of likely behavioural responses from the rival firm. Further regression models were also estimated taking changes in consumer surplus and economic welfare as dependent variables. These functions facilitate analysis of the impact of competition on passenger benefit and overall welfare changes over different competitive strategies. The specific results of this exercise are commercially sensitive and cannot be reproduced in detail here. However, it is worthwhile comparing the principal findings with those of our inter-urban case study. The best non-collusive strategies available to the entrant and the incumbent were to discount fares and produce at a high level of output. This scenario results in a situation where both operators make a loss and although consumers benefit greatly, overall welfare is reduced substantially. This reduction in welfare arises from the fact that reductions in operator profitability are on average five times greater than increases in consumer surplus. As with the analysis of competition on the inter-urban route, we concluded that head-on competition would be unattractive for the entrant and for social welfare. We concluded that the market was well served by the incumbent and that there was insufficient scope for the market to grow significantly so as to generate sufficient revenue to cover the costs of new entry. Competition is therefore likely to be damaging because it reduces load factors and raises unit cost. Unless the rail market was more responsive to reductions in fares and increases in the level of service, head-on competition is likely to be welfare negative. Indeed, sensitivity tests indicate that, in this case, absolute fare and generalised time elasticities greater that 2.0 are required to generate welfare positive outcomes.

The above results assume that competition does not have any significant effect on costs. The experience of industries such as refuse collection, hospital domestic services and local bus services (Domberger et al., 1986, 1987, Heseltine and Silcock, 1990) suggests that competition can lead to a reduction in average operating costs of 30%. If this were the case for rail also, then it would outweigh the loss of consumer surplus in many of the scenarios studied. However much of the cost reduction in these industries was due to competitive tendering and arguably the franchising process will have already led to these reductions in rail costs. Evidence from Sweden suggests that tendering rail services has reduced costs by 15-10%, whilst the 'privatisation' of JNR in Japan has reduced operating costs by over 30% (van de Velde et al., 1998).

Overall Review

The experience of many previous attempts to inject competition into an industry that has been characterised by monopoly is that there is a mixture of expected and unexpected results. The reform of railways in Britain is no exception. Table 10 presents some data on the performance of the British rail industry since the 1993 Railways Act (based on DETR, 1997, 1998c). We also present some lessons that we believe can be learnt from the British experience so far.

We should start by emphasising that the reform process is only at the halfway stage. The industry has been successfully privatised, with the main sell-offs and franchising occurring in 1996 and adjacent years. However, the second half will only get under way in around 2003 when the first round of tenders are due for renewal, the ROSCO contracts begin to expire and open access competition for passenger railways is possibly permitted.

Those evaluations of privatisation that have been undertaken (e.g. Harris and Godward, 1997, White, 1998) have taken place at half time rather than at the end of the match. Moreover, the results of such evaluations are affected by the choice of the base year (e.g. 1993/4 or 1995/6), assumptions concerning the counterfactual (i.e. what would have happened if the reforms had not taken place) and assumptions concerning transfers.

Furthermore, on-the-track competition has so far been limited to where existing franchises overlap or where railway geography permits route competition. Additional on-track competition is likely in the future with the introduction of the second stage of the moderation of competition in September 1999. The extent to which this competition will be beneficial will be largely determined by how innovative the new services provided are and/or the extent to which they lead to cost reductions. It is possible, that by diluting load factors, competition may in certain circumstances lead to increases in average costs.

Off-the-track competition has been substantial. There have been typically five serious bids per franchise: with one bid from a management buyout (MBO) team; two from bus companies and one from a company with other transport concerns. There have usually been a similar number of bids that were not deemed serious. The competition for franchises appears to have become more serious in later rounds, at least in terms of subsidy required. Those who entered the process early seemed more likely to earn profits than those who entered late. Similar trends of higher prices in later rounds have been noted in other privatisations e.g. of the National Bus Company. A phased approach to introducing off-track competition therefore seems sensible.

Fifteen out of the 25 franchises have been won by bus companies. Off-the-track competition was highly dependent on the existence in Britain of a deregulated and privatised bus and coach industry that was competing in the same markets as rail. The existence of intermodal competition has been important in promoting intramodal competition, although there have been instances where there may be trade-offs between the two types of competition. For example, National Express has been forced by the MMC to divest of Scottish CityLink so as to permit their take-over of ScotRail.

Four alliances already account for 70% of the passenger rail business by income, whilst one operator dominated the freight market. There have thus been very strong tendencies towards re-agglomeration. This may have implications for the dynamics of franchising as competition may be less intense in the second and subsequent rounds. Other issues likely to emerge are the Government's response to business failures (which given our analysis in Table 3 are likely), the extent of incumbency advantages in contract renegotiations, the impact of 'passenger dividends' and the ease of asset handovers. Our bidding model is a form of independent value auction in which the winning bid increases (i.e. subsidy decreases) as the number of bidders increase. This would suggest that maintaining the level of off-track competition should be an important policy priority.

Table 10   Some initial effects of the reforms of the British railway system

	1993/4	1994/5	1995/6	1996/7	1997/8
Loaded train km (m)	350.2	340.2	353.5	360.0	N/A
Passenger km (b)	30.4	28.7	30.0	32.2	34.2
Revenue per passenger km (pence - 1997/8 prices)	7.9	8.2	8.4	8.2	8.3
Freight tonne km (b)	13.8	13.0	13.3	15.1	16.9
Revenue support (£m 1997/8 prices)	1201	2161	2189	2153	1804
Staff (thousands)	135.2	133.5	130.6	122.1	N/A
Punctuality	90.3	89.6	89.5	92.5	90.4
Reliability	98.8	98.8	98.8	99.1	98.9

N/A Not available

Table 10 shows that total government support for the first year of the new regime (1994/5) was £2161 million (in 1997/8 prices), compared to the previous year's £1201 million. This increase is largely due to higher capital costs for infrastructure and rolling stock. However, subsidy is forecast to fall sharply. By 1997/8 Government support was £1804 million, a decrease of almost 17% on the 1994/5 figure. In the first year of private sector operations, only one franchise did not require subsidy. By the final year, this is forecast to increase to 10 (OPRAF, 1997). Off-track competition does seem to be improving financial performance.

As far as passengers are concerned, the level of service in terms of train kms has increased between 1993/4 and 1996/7 by 3%. Between 1993/4 and 1997/8 passenger kms have increased by over 12% and fares, or more correctly yields, have increased (despite regulation from OPRAF) in real terms by around 5%. There is some evidence to suggest that growth has been concentrated on London bound intercity and commuter markets (Root and Preston, 1998). Between 1994 and 1998 GDP has grown in real terms by around 10%. Conventional wisdom concerning rail GDP elasticities suggests that a figure of around or slightly above unity is appropriate. A substantial element of the demand growth may therefore be attributed to the economic up-turn. It has been suggested that some of the rest of the increase in patronage figures is due to better revenue control.

Punctuality increased by 1.9 percentage points and reliability by 0.3 of a percentage point between 1993/4 and 1996/7. There has since been some slippage so that in 1997/8 both punctuality and reliability are only 0.1 of a percentage point higher than in 1993/4. The cause of this deterioration is thought to be due to the problems the TOCs have had in carrying extra passengers and running more services. The proportion of delays attributable to Railtrack appears to have declined. Passenger complaints have gone up markedly but this may partly be due to the high public expectations of the advantages of the privatisation process and better reporting procedures.

Labour productivity has improved dramatically since the reform process started. Assuming only a modest reduction in the overall workforce (Table 10 suggests a decline of only 10% but there are severe problems about the treatment of contractors etc), the number of passenger kilometres per employee has increased by around 24% between 1993/4 and 1996/7, whilst the corresponding figure for train kilometres per employee is estimated at 14%.

There have been some signs of increased innovation in the industry. In part, these are related to franchise commitments particularly to provide more service, new and refurbished rolling stock and improved stations. However, there have been important commercial initiatives particularly in the areas of ticketing (e.g. group travel, bonus schemes) and in retail and distribution (e.g. the creation of telesales centres).

There has been a mini-revival in the rail freight industry. In the two years of private operation, tonne kms has increased from 13.3 billion in 1995/6 to 16.9 billion in 1997/8 (up 27%). Some of this growth is due to the Channel Tunnel. Although intramodal rail freight competition is limited, there is intense intermodal competition (which is likely to intensify with the introduction of 40 tonne lorries in January 1999). Competition in the final product market (e.g. if rail charges too much for coal, power generators can switch to gas) may also have some impact.

Our conclusions concerning the rolling stock and infrastructure businesses are circumspect. More detailed analysis is required. The rolling stock leasing market does not seem to be competitive, with the three ROSCOs reluctant to compete with each other. However, there is every expectation that the market will be competitive in the future, particularly as other players such as the rolling stock manufacturers become involved. The lack of regulation of the ROSCOs seems to have been an oversight as was the lack of claw back provisions for future sales given that all three ROSCOs have been sold on for large profits (National Audit Office, 1998). ORR (1998b) has indicated the need for a concordat between the Regulator and the ROSCOs but it is not clear how this will work in practice.

Railtrack's performance appears mixed. On the one hand, the incentive regime appears to have had a marked effect on performance. The recent deterioration in reliability and punctuality does not seem to be the fault of Railtrack. On the other hand, there have been severe problems in devising a transparent access and charging regime and in encouraging investment. A review of access charging is due in 2001, following the earlier critical review in 1995 (ORR, 1995). Railtrack's Network Management Statement has been heavily criticised by the Regulator (ORR, 1998c). The regulation of Railtrack remains a difficult problem with both the Rail Regulator and the Franchising Director (and in the future the Strategic Rail Authority) having a role to play. The infrastructure services supply industry appears to be competitive and is likely to be more so in the future as Railtrack's long term contracts with its traditional suppliers come to an end. A possible way of introducing competitive pressure on Railtrack would be an unbundling by splitting the company up into its zonal components (akin to AT&T's Baby Bells). This would then allow some yardstick competition (see Schleifer, 1985) and might, in the long run, provide a market test for vertical integration. An alternative would be to further incentivise Railtrack. To an extent, this has been done with respect to the profit sharing agreement Railtrack has made with Virgin concerning the West Coast Upgrade, although this particular arrangement is not without its regulatory difficulties.

Overall, the reform process has been incredibly complex and hence costly. The National Audit Office (1996) estimated the cost at greater than £1 billion. Harris and Godward (1997) put the figure at £5 billion. There is a need for more detailed calculation of these transitional costs. Nonetheless, the process has been made to work. A key issue remains whether alternative regimes such as vertically integrated concessions would have worked better.

Conclusions

Our work has suggested that off-track competition can reduce subsidy for most franchises, whilst maintaining current services and fare levels and is thus likely to be welfare positive. Larger franchises, looser regulation and protection from competition will all reduce subsidies (by as much as 25% on average) although they may have other disadvantages. Further subsidy reductions can be achieved, but they may be at the expense of fare increases and service reductions, with uncertain welfare implications. We conclude that off-track competition may be particularly effective in promoting productive efficiency. The impact on allocative and dynamic efficiency is less clear but is likely to be positive.

Our work has suggested that the most likely form of on-track competition is cream skimming. This can increase benefits to users but reduces welfare because of reductions in producer surpluses. There may be some instances where on-track competition leads to benefits due to innovative pricing and/or services but they do not feature in the tests we have run. We conclude that on-track competition is likely to be welfare negative unless it is very carefully regulated to prevent cream skimming behaviour. Moreover, the interaction with off-track competition is likely to lead to higher subsidy requirements. We conclude that on-track competition can lead to allocative inefficiencies, particularly with respect to network benefits such as interavailable ticketing. These disadvantages may be offset by gains in productive and dynamic efficiency. A case by case approach may be sensible.

As for the future, subsidies are expected to decrease for all franchises. Some of them are expected to turn into 'profits' within the current franchising period. If this is realised, the second half of the reform process may see the appearance of a two tier network, as envisaged by, for example Foster, 1994. A substantial part of the network might then be operated commercially, under open access provisions, whilst the unprofitable services would be franchised. The passenger rail market would thus become like the local bus market.

The process of concentration of the passenger business into a few large alliances may continue in the future. This would provide a regulatory dilemma in that concentration may be desirable from a cost-efficiency viewpoint but may lead to franchising becoming less competitive in the future. To some extent these problems may be tempered by permitting a greater degree of on-the-track competition from September 1999.

The 1998 White Paper's proposals to create a Strategic Rail Authority (SRA) and re-define the Rail Regulator's role will address some of the problems that have emerged (DETR 1998a and b). An SRA concentrating on consumer interests at the operating level and an ORR focusing on the regulation of Railtrack and the ROSCOs and on the administration of competition policy seems sensible. However, a legislative timetable has yet to be established and it is unlikely that the proposals will come into force before 2001. Moreover, there still appears to be likely duplication of effort between the SRA and ORR, particularly concerning Railtrack's investment levels. The effective regulation of Railtrack thus remains unresolved, despite the windfall tax imposed in the July 1997 budget. The Comprehensive Spending Review proposes to increase expenditure on public transport by £1.7 billion over the period 1999 to 2002 (HM Treasury, 1998), and it is likely that some of this will be absorbed by the Rail Passenger Partnership programme and the Infrastructure Investment Fund proposed by the White Paper. However, it is unlikely to be sufficient to overcome the historic investment shortfall.

We conclude that off-track competition has been effective for passenger operations but on-track competition is likely to be less effective and less compatible with an integrated transport policy. Intermodal competition, competition between tracks and competition on the train (i.e. between different standards of service) may be effective in certain instances. We believe that on-track competition may be more effective for the rail freight industry (or at least the threat of competition), although in practice intermodal competition has a greater impact. The supply of infrastructure services and rolling stock will become more competitive (and effective) in the future. The unresolved issue is how to ensure the level and price of infrastructure approximates competitive levels. We are not convinced that either the simple auctioning schemes of Starkie (1993) or the more complex schemes of Nilsson (1995) are practical. Yardstick competition may represent a possible way forward.

References

BRB (British Railways Board) (1994) *Annual Report and Accounts* BRB, London.

Chadwick, E. (1859) On Different Principles of Legislation and Administration *Journal of the Royal Statistical Society*, 22, pp.381-420.

Demsetz, H. (1968) Why Regulate Utilities? *Journal of Law and Economics 11*, 55-65.

Department of the Environment, Transport and the Regions (DETR) (1997) *Transport Statistics Great Britain 1997.*

Department of the Environment, Transport and the Regions (DETR) (1998a) *A New Deal for Transport: Better for Everyone.* Cm3950. The Stationery Office, London.

Department of the Environment, Transport and the Regions (DETR) (1998b) *Railways Policy: a Response to the Third Report of the Environment, Transport and Regional Affairs Committee on the Proposed Strategic Rail Authority and Railway Regulation* Cm 4024. The Stationery Office, London.

Department of Environment, Transport and the Regions (1998c) *Bulletin of Rail Statistics Quarter* 1 1998/9. Government Statistical Service, London.

Dodgson, J.S. (1994) Access Pricing in the Railway System *Utilities Policy*, 4, 205-213.

Domberger, S., Meadowcroft, S. and Thompson, D. (1986) Competitive Tendering and Efficiency: The Case of Refuse Collection. *Fiscal Studies.*

Domberger, S., Meadowcroft, S. and Thompson, D. (1987) The Impact of Competitive Tendering on the Costs of Hospital Domestic Services. *Fiscal Studies.*

Foster, C. (1994) *The Economics of Rail Privatisation.* Centre for Regulatory Studies, CIPFA, London.

Galvez, T. (1989) *Operating Strategies for Rail Transport.* PhD Thesis, School of Economic Studies, University of Leeds.

Harris, N (1997) *Analysis of the Subsidies Bid for British Rail Passenger Franchises.* PTRC European Transport Forum, Brunel University.

Harris, N.G. and Godward, E. (1997) *The Privatisation of British Rail.* The Railway Consultancy Press, London.

Heseltine, P.M. and Silcock, D.T. (1990) The Effects of Bus Deregulation on Costs. *Journal of Transport Economics and Policy*, 24, 3, 239-254.

HM Treasury (1998) *Modern Public Services for Britain: Investing in Reform.* Cm 3798, The Stationery Office, London.

Nash, C.A. (1995) *Rail Privatisation -The Experience So Far.* Paper presented to 4th International Conference on Competition and Ownership in Land Passenger Transport. Rotorua, New Zealand.

Nash, C.A. (1996) The Separation of Operations from Infrastructure in the Provision of Railway Services. *Round Table* 103, ECMT, Paris.

Nash, C.A. (1997) *Rail Privatisation - How Is it Going?* Presented to the CIT, Yorkshire and Humberside Region.

National Audit Office (1996) *The Award of the First Three Passenger Rail Franchises.* Report HC 701, HMSO, London.

National Audit Office (1998) *Privatisation of the Rolling Stock Leasing Companies.* HC 576 Session 1997-98. The Stationery Office, London.

Nilsson, J-E (1995) *Allocation of Track Capacity. Experimental Evidence on the Use of Vickrey Auctioning in the Railway Industry.* CTS Working paper, 1995:1, Borlange, Sweden.

Office of Passenger Rail Franchising (1997) *Annual Report 1996/7*. OPRAF, London.

Office of the Rail Regulator (1994) *Competition for Railway Passenger Services. A Policy Statement*. ORR, London, December.

Office of the Rail Regulator (1995) *Framework for the Approval of Railtrack's Access Charges for Freight Services: A Policy Statement*. ORR, London, February.

Office of the Rail Regulator (1998a) *New Service Opportunities for Passengers: Criteria and Procedures for the Approval of Train Operators Proposals for Stage II of Moderation of Competition*. ORR, London, March.

Office of the Rail Regulator (1998b) *Review of the Rolling Stock Market*. ORR, London, February.

Office of the Rail Regulator (1998c) *Rail Regulator's Review of Railtrack's 1998 Network Management Statement*. ORR, London, July.

Preston, J.M. and Whelan, G.A. (1995) *The Franchising of Passenger Rail Services in Britain*. Presented to the 4th International Conference on Competition and Ownership in Land Passenger Transport, Rotorua, New Zealand.

Preston, J.M. and Whelan, G.A. (1996) *The Sale of the Century? An Analysis of the Franchising of British Rail*. Presented to the 28th Annual UTSG Conference. University of Huddersfield.

Preston, J., Whelan, G., Nash, C. and Wardman, M. (1997) The Franchising of Passenger Rail Services in Britain. Submitted to *International Review of Applied Economics*.

Preston, J., Whelan, G. and Wardman, M. (1998) An Analysis of the Potential for On-track Competition in the British Passenger Rail Industry. Submitted to the *Journal of Transport Economics and Policy*.

Rail Operational Research (1997) *The Passenger Demand Forecasting Handbook*. London

Root, A. and Preston, J. (1998) *Research on Railway Competition - National Report Great Britain*. TSU Working Paper 861, University of Oxford.

Schleifer, A. (1985) A Theory of Yardstick Competition. *Rand Journal of Economics*, 16, 3, 319-327.

Starkie, D. (1993) Train Service Co-ordination in a Competitive Market. *Fiscal Studies*, 14, 2, 53-64.

Van de Velde, D., Mizutani, F., Preston, J. and Hulten, S. (1998) *Railway Reform and Entrepreneurship: A Tale of Three Countries*. European Transport Conference, University of Loughborough.

Whelan, G.A., Preston, J.M, Wardman, M.R. and Nash, C.A. (1997B) *The Privatisation of Passenger Rail Services in Britain: An Assessment of the Impacts of On-the-Track Competition*. Presented to the European Transport Forum. PTRC, London.

Whelan, G.A., Wardman, M.R., Preston, J.M and Nash, C.A. (1997A) *Rail Privatisation. The Development of a Disaggregate Demand Model*. Technical Note 404, Institute for Transport Studies, University of Leeds.

White, P. (1998) *Financial Outcomes of Rail Privatisation in Britain.* European Transport Forum, Loughborough University.

Willich, A. (1996) *Rail-links in Hampshire.* MSc Dissertation, University of Oxford

Worsey, S.J. (1994) *Alternative Organisational Arrangements for Rail Transport.* PhD Thesis, School of Business and Economic Studies, University of Leeds.

Note

1   This work was partly undertaken as an Economic and Social Research Council Project entitled *The Privatisation of Passenger Rail Services: Analysis and Monitoring* [R000234735]. This chapter draws heavily on an earlier paper which was presented as 'The Privatisation of Passenger Rail Services in Great Britain: An Evaluation of Competition for the Market and Competition in the Market' to the World Conference on Transport Research in Antwerp on 14 July 1998. The lead author of that paper was Gerard Whelan and the co-authors were Mark Wardman, Chris Nash and myself. The conclusions in this paper and any errors therein are purely my own.

7 Effective Competition in the Bus Industry

JOHN DODGSON

Introduction: Buses and The White Paper

> Deregulation of the local bus market, outside London, caused substantial upheaval because of 'bus wars' and confusion over changing service patterns. There have been some good examples of innovation but frequent changes to bus services, poor connections and the reluctance of some bus operators to participate in information schemes or through-ticketing undermined bus services. In this climate, it was not easy for buses to match the levels of comfort, reliability and access offered by the private car.
>
> Deregulation has not broken the spiral of decline in local bus use. Since 1986 bus use has fallen by about a quarter - by about one billion fewer journeys a year; in contrast with London, within a regulated market, where use has held up. More recently, there have been good examples of bus companies and local authorities working together in Quality Partnerships to change the image of bus services and stem, sometimes even reverse, the decline in patronage.
>
> (DETR, 1998, p.28.)

In this way, the Integrated Transport White Paper characterises post-deregulation experience in the bus industry. However, the White Paper does not propose to re-regulate the industry, but rather to seek to improve the quality of bus services by a number of measures, particularly through increased use of Quality Partnerships, and possibly Quality Contracts, involving both bus operators and local authorities[1]. The White Paper stresses measures to improve bus service quality through:

- more bus lanes and other improvements to bus priority;
- improved passenger information systems, including co-ordinated timetable information;
- measures to reduce timetable instability;
- better design of buses, including improvements to permit disabled access;

- better physical interchange between bus services, and between buses and other public transport modes; and
- improved safety at bus stops.

The bus industry will therefore remain a competitive one. The main aim of this paper is to assess how effective this competition has been, and is likely to continue to be.

Effective Competition

Hence the title of the paper is 'Effective Competition in the Bus Industry'. I wish to consider whether the types of competition resulting from local bus deregulation have been successful, 'effective'. It is first necessary to consider what might be meant by 'effective'. There are a number of possible meanings:

- Effective in ensuring an economically-efficient outcome in which prices equal marginal costs and dynamic efficiency is achieved in the sense that there is optimal innovation in the provision of new types of service.
- Effective in reversing the long-term decline in local bus patronage.
- Effective in minimising the costs of whatever service levels are provided.
- Effective in maximising operators' profits.
- Effective in ensuring that bus passengers are not exploited.

These are, of course, to some extent mutually exclusive.

To consider whether competition has been effective, it is next necessary to review what we now know about competition in bus markets.

Competition in Theory and Practice

Contestability?

> The bus market is therefore a highly contestable one.
> (Department of Transport, 1984, p.52)

Before deregulation it might have been possible to caricature alternative views about the likely impact of bus deregulation into two camps. One view would be that competition can solve everything, and ensure static and dynamic efficiency in the market. The opposite view was that 'on-the-road' competition is essentially 'wasteful', so that deregulation would lead back to the bad old days of the 1920s.

The first view would have been correct if the bus market had turned out to be perfectly contestable. In fact, if any industry was ever going to be contestable, surely this was it? The minimum size of operation was one bus, there was a thriving second-hand market for buses, no secrets in how to operate buses, no post-deregulation restrictions on entry to the market, and plenty of econometric evidence of constant returns to scale with regard to fleet size.

However, one requirement for contestability was missing, namely the possibility of 'hit-and-run' entry. The essential prediction of contestability is that the mere possibility of entry forces firms to set existing prices at average costs, and forces them to incur average costs that are minimised for the output actually produced. In practice existing bus operators need only reduce prices to average costs once entry occurs, or is clearly on its way. However, they can only effectively meet competition if their costs are at, or reasonably close to, the allocatively efficient level.

The other evidence that the bus market is not perfectly contestable is the observation of competition agency activity. If a market is perfectly contestable then the economically-efficient outcome will be guaranteed, and competition agencies need play no role. There will be no scope for firms to engage in predatory behaviour, and no need to investigate mergers because in a contestable market optimal industry structure will be determined endogenously. Given the massive role played by the Office of Fair Trading and the Monopolies and Mergers Commission in the bus industry since deregulation we can conclude that the industry is certainly not perfectly contestable - unless the UK competition policy agencies have been singularly misguided for the last twelve years.

If the market is not perfectly contestable, then existing firms have some market power, and can charge prices that to some extent exceed marginal costs.

Fare Competition

     Passengers board the first bus.
     (Conventional wisdom)

Fare competition has been rare. There have been some examples of entrants deliberately charging lower fares than existing operators. For example, Fareways in Liverpool was a good example of an independent company that set up to operate in low income areas and deliberately charged lower fares on all their services. However, Fareways eventually raised fares, and were subsequently bought out by the main local operator. The White Paper (DETR, 1998, p.43) quotes the current example of

Stagecoach's Magicbus operation in South Manchester[2]. However, most operators realised that one of the main features of the industry is that 'passengers board the first bus', and there is little extra revenue to be gained, and plenty to be lost, by charging lower fares than established operators on local services.

Product Differentiation/Quality Competition

> When I asked her where to catch a bus to the station, the bemused shopper chortled: 'Don't ask me, love. There are millions of buses in this town and most of them are empty anyway. I just walk everywhere'.
> This is Darlington...
> (The Guardian, December 1994)

Like most real-world markets, the bus industry is one characterised by product differentiation. Economists have used two main types of model to consider product differentiation, models of horizontal product differentiation and models of vertical differentiation (Dodgson and Katsoulacos, 1988a). With vertical product differentiation all consumers will rank different 'brands' in the same order of preference if they are all sold at the same price, i.e. everyone agrees which is the best brand, which the second-best, and so on. With horizontal product differentiation consumers will choose different brands even if they are all available at the same price. In practice, models of horizontal product differentiation have turned out to be much more relevant for the bus industry than models of vertical product differentiation.

The primary possible example of vertical differentiation in local bus markets was that between slower conventional double-deck buses, and faster minibuses. All consumers prefer the faster service, but with vertical product differentiation would expect to have to pay a higher fare to use it - since the slower service could only survive in the market place by offering lower fares. In Dodgson and Katsoulacos (1988b) we showed how a model of vertical product differentiation developed by Shaked and Sutton (1982) could be applied, at least in theory, in the bus industry. This model showed that under a wide range of conditions a market could sustain at most two levels of product variety.

At the end of the paper we noted that local public transport markets already contained two price-quality combinations, namely low-fare/low-quality buses, and high-fare/high-quality taxis, and that there might not be room for differentiation between different price-quality combinations of buses. Admittedly, this was some rather ad hoc speculation tacked on the end of some quite complex theory, but events may have proved it correct. The ability of operators to provide a high quality bus alternative, especially in crowded streets, may be very limited, and certainly we believe that it has been rare for such to be observed in practice.

On the other hand, horizontal product differentiation is common, because:

- Potential bus passengers wish to travel on particular routes, and most routes are not substitutes for each other; and
- Even on particular routes, potential passengers have different preferred times of travel.

This has led to widespread use of Hotelling-type location models to analyse competition in bus markets. These models often predict instability in service patterns, with clustering of services, and frequent changes in timetables, though the models are less good at predicting the numbers of operators serving the market as opposed to the numbers of services.

These and other models provide an explanation of the instability so often observed in bus markets. In addition Alison Oldale (1998) has recently used a game-theoretic framework to explain why timetable instability may be common. This means that competition can be wasteful, since a regular headway service with a given number of buses per hour should cost the same to operate as a bunched service with the same number of buses per hour, but will provide greater consumer benefits if potential travellers' optimal departure times are evenly spread around the hour. The 'curious old practices' identified by Foster and Golay (1986) do have a rational explanation. In turn, this means that one of the objectives of government intervention may be achievement of what may be termed an 'orderly market'.

Predatory Behaviour

> We find these actions to be predatory, deplorable and against the public interest.
> (Monopolies and Mergers Commission, 1995, p.3, on Busway's behaviour in Darlington)

Another major feature of bus markets has been the emergence of predatory behaviour as centre stage in competition policy investigations. Before 1986 predatory behaviour had been dismissed by some industrial economists, who regarded it as irrational and therefore (since everyone, including potential prey, knew it to be irrational) not likely to occur. Nevertheless, a growing literature had used more complex notions of asymmetric information to argue that predatory behaviour could be rational in certain circumstances, including situations of multi-market competition.

Naturally bus operators had better things to do than keep abreast with this literature, but fortunately the Office of Fair Trading did not. The UK bus industry literature is a fertile ground for studies of predatory behaviour, both in OFT reports, and in a number of the MMC reports on mergers and/or competition, and in particular the 1995 report on the north-east which investigated five separate instances of predatory behaviour.

One of the features of predatory behaviour is that it is difficult to detect, especially where it is outlawed. Some conclusions that have emerged are:

- It is very difficult for an incumbent to know what to do when faced by an unsophisticated entrant. This was one reason why we thought that the OFT (1989) and MMC (1990) reports were somewhat one-sided in dealing with Highland Scottish Omnibuses in Inverness (Dodgson and Katsoulacos, 1993);
- While incumbents might prey on entrants, it is possible for entrants to prey on incumbents. The Bognor Regis investigation (OFT, 1992) is a case in point;
- Given the absence of particularly serious penalties in many cases[3], a reputation for predatory behaviour which has been confirmed by the OFT or MMC might be of positive benefit in the future;
- It is possible that small firms might make a nuisance of themselves in order to increase their sale value to a larger competitor (MMC report on the North East, 1995).

In our own work on the investigation of predatory behaviour in the bus industry, we argued following Phlips (1995) that predatory behaviour should be regarded as action which turned a profitable entry opportunity into a negative one. This involved a modelling framework which identified the counter-factual, and considered the actions of both the incumbent and the entrant. Although such models will be dependent on specific functional forms used for demand and cost functions, we were able to show how they more clearly identified alternative explanations of what might have happened in particular cases. The OFT subsequently used our model in a number of its later predatory behaviour investigations: the Southend report (OFT, 1993) provides the fullest explanation.

The MMC have characterised their approach to merger rulings as that of a 'patchwork quilt' policy of trying to limit the geographical monopolies of individual major operators (MMC, 1995, p.33). Associated with this is the idea of the avoidance of 'safe havens', areas so large that they cannot be penetrated by powerful neighbours. Of course, competition agencies cannot get the correct balance every time, as evidenced by the end-1997 decision to require divestment of some FirstBus routes in Glasgow following their take-over of SB Holdings, on the grounds that the Greater Glasgow area would otherwise provide a safe haven for FirstBus, a prediction that was proved false by events almost before the ink on the MMC's report was dry. Despite this, I believe that on balance the involvement of the competition agencies in the British bus industry has had a positive impact, though there will have been administrative and compliance costs which would be interesting to quantify.

The Results of Competition

We turn now to look at the results of competition, as evidenced from annual statistics on local bus services in England[4] published in the DETR's Bus and Coach Statistics Great Britain 1996/97 (DETR, 1997).

Figure 4.1 shows passenger journeys on local services in London, the six Mets, and the Shire counties between 1985/86 and 1996/97. This shows the relatively constant traffic levels in London, and the declines in the Mets and Shires. By 1996/97 the three segments of the market accounted for roughly similar numbers of passengers.

As is well-known, this decline in traffic has been accompanied by growth in capacity, as measured in bus-kms. Figure 4.2 shows the pattern in England since deregulation. Growth was most rapid in the period immediately after deregulation, and in the Shire counties, associated in particular with the expansion of minibus services. However, growth has continued despite the declines in patronage, with accelerating growth of capacity in the Shire counties in the last few years.

Results in terms of passenger journeys per bus-km are shown in Figure 4.3. This is a somewhat unsatisfactory measure of capacity utilisation because it does not take account of differences in average journey length between sectors, changes in average journey length within sectors over time, and changes in average vehicle capacity. Nevertheless, the figure shows the significant declines in this measure of capacity utilisation in Shires (down 42 per cent) and Mets (down 50 per cent), together with the decline and then some recovery in London.

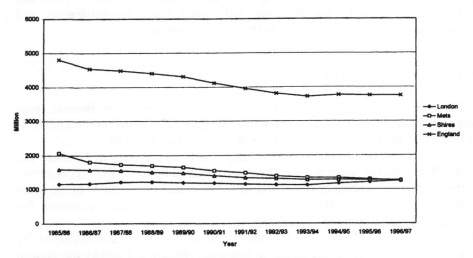

Figure 4.1   Local bus passenger journeys in England

Figures 4.4 and 4.5 show what has happened to fares in real terms. Figure 4.4 shows the DETR's local bus fare indices for England since deregulation, with 1985/86 fare levels in each type of area normalised to 100.   Growth was most rapid in the Mets in the year following deregulation, mainly as a result of the reversal of the low fares policies in South Yorkshire and Merseyside. Taking the period since 1986/87, the fare index shows a rise of 29% in the Mets, 12% in the Shires, and 34% in London.

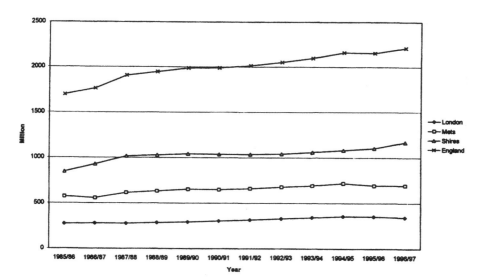

Figure 4.2   Local bus kms in England

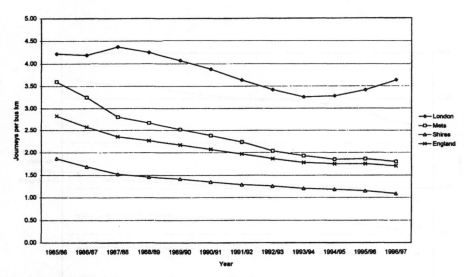

Figure 4.3   Passenger journeys per bus km

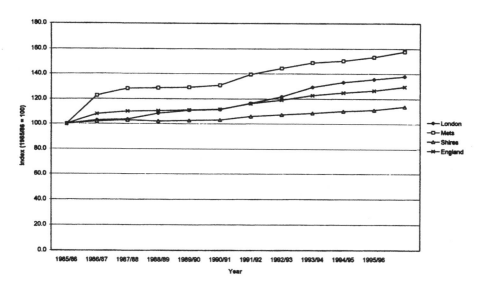

Figure 4.4   Local bus fare indices in England (1985/86 = 100)

Figure 4.5 shows actual fare levels at 1996/97 prices in the form of average fare per passenger journey. Differences between types of area will partly reflect differences in average journey length, and trends could reflect changes in average journey lengths within sectors of the market. Between 1986/87 and 1996/97 this measure of average fare shows increases of 36 per cent in the Mets, 17% in the Shires, and 15% in London.

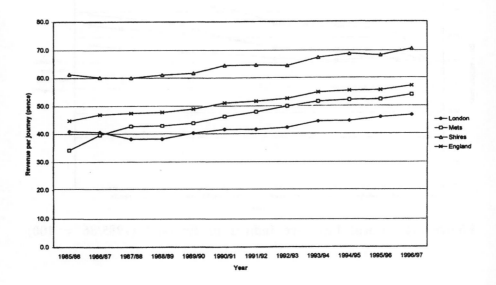

Figure 4.5   Local bus revenue per journey at 1996/97 prices

Figure 4.6 shows operating costs, including depreciation, per bus km. Operating costs per bus km are highest in London, and lowest in the Shires. Since 1985/86 operating costs per km have fallen by 49% in the Mets, 42% in the Shires, and by 45% in London.

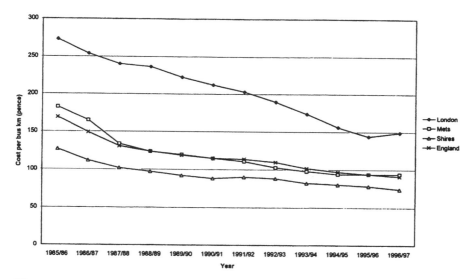

Figure 4.6 Operating cost (including depreciation) per bus km at 1996/97 prices

Finally Figure 4.7 combines data on operating costs and patronage to show operating costs, including depreciation, per passenger journey. While costs per journey have fallen in London as patronage has held up, in the Mets and the Shires the impact of declining patronage has generally outweighed the impact of reduced costs per bus km.

How Effective Has Competition Been?

We return finally to the main theme of the paper, namely that of the effectiveness of competition in the industry.

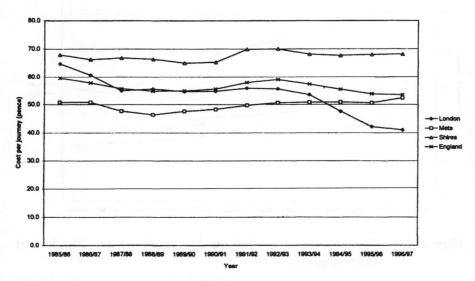

Figure 4.7 Cost per passenger journey (including depreciation) at 1996/97 prices

Effective in Ensuring an Economically-Efficient Outcome?

There is no doubt that the bus industry is not perfectly contestable. In particular, the threat of hit-and-run entry is weakened because of the ability of incumbents to react to entry when it occurs. Nevertheless, incumbents' ability to do this must be partly conditional on their being able to compete on cost levels with entrants: in other words, given the nature of the industry, there is little room for firms whose costs are inflated by inefficiencies.

It is also not clear that the market has provided the optimal mix of price and service quality. While a criticism of the regulated regime was that bus networks did not evolve to meet changing consumer demands, it is doubtful whether consumers believe that the market has provided the optimal rate of service pattern and frequency adjustment. While it could be argued that rising incomes would lead to demand for a higher quality/higher price combination than in the past, I am not convinced either that the patterns of fare and bus-km combinations revealed in the Mets and the Shires by combining figures 4.4 and 4.2 are the economically-efficient ones, though it would be very interesting to see a modelling framework that did reveal the optimal combination over time.

Effective in Reversing the Long-term Decline in Local BusPatronage?

This would have been a tall order, and it is unlikely that it could ever have been achieved. As figure 4.1 showed, deregulation has not led to a decline in the provision of bus services. What might be questioned is whether deregulation stemmed the decline as much as it might have. At first sight the evidence from London might suggest it did not, because in un-deregulated London bus patronage has held up, while costs have fallen with competitive tendering[5].   However, NERA's 1997 study showed that there were particular demographic circumstances in London favouring bus travel. In particular, car ownership fell by around one per cent per head per year in London in the 1990s, whereas in the rest of the country it rose by around 1.4% per head each year. In addition, population density in London rose by 3.5% in London in the decade from 1985 to 1995, while in the Mets it fell by 1% (NERA, 1997, pp.13-14).

Effective in Minimising the Costs of Whatever Service Levels are Provided?

Whatever form of competition is involved, whether it be competition through tendering, or competition in service provision the market place, competition has had a major impact in reducing operating costs. In terms of productive efficiency competition can be judged to be effective.

Effective in Maximising Operators' Profits?

Here again we might judge competition to be effective. Despite the decline in patronage, the profitability record of the industry has not been bad. Of course, there has been a lot of 'buying and selling', which may indicate that an advantage of deregulation and privatisation was that it enabled a clearer view to be taken of the relative value of different types of assets, including real estate assets. In addition, I would suggest that the private sector may be in a better position to take a realistic view of longer term market prospects than the public sector, which is sometimes prone to a certain amount of wishful thinking when it comes to assessing the future of public transport.

Effective in Ensuring that Bus Passengers are not Exploited?

If the answer to this question was 'yes', then we could dispense with the OFT and the MMC so far as bus industry matters are concerned. I have already indicated that I think that competition agencies do have an important role to play in the industry.

References

Department of Transport, Scottish Office, Welsh Office (1984) *Buses* Command 9300, HMSO, London.

DETR (1997) *Bus and Coach Statistics Great Britain (1996/97)* The Stationery Office, London.

DETR (1998) *A New Deal for Transport: Better for Everyone.* The Stationery Office, London.

Dodgson, J. S. and Katsoulacos, Y. (1988a) Models of Competition and the Effect of Bus Service Deregulation In J. S. Dodgson and N. Topham (eds) *Bus Deregulation and Privatisation: an International Perspective.* Avebury, Aldershot, 45-68.

Dodgson, J. S. and Katsoulacos, Y (1988b) Quality Competition in Bus Services: Some Welfare Implications of Bus Deregulation *Journal of Transport Economics and Policy,* 22, 263-281.

Dodgson, J. S. and Katsoulacos, Y. (1991) Competition, Contestability and Predation: the Economics of Competition in Deregulated Bus Markets *Transportation Planning and Technology,* 15, 263-275.

Dodgson, J. S., Katsoulacos, Y. and Newton, C.R. (1994) An Application of the Economic Modelling Approach to the Investigation of Predation *Journal of Transport Economics and Policy,* 27, 153-170. Reprinted in Tae Hoon Oum et al (1997) *Transport Economics: Selected Readings.* Harwood Academic Publishers, Amsterdam, 539-563.

Foster, C. D. and Golay, J (1986) Some Curious Old Practices and Their Relevance to Equilibrium in Bus Competition *Journal of Transport Economics and Policy,* 20, 191-216.

Kennedy, D. (1995) London Bus Tendering: the Impact on Cost *International Review of Applied Economics,* 9, 57-63.

Monopolies and Mergers Commission (1990) *Highland Scottish Omnibuses Ltd* HMSO, London.

Monopolies and Mergers Commission (1995) *The Supply of Bus Services in the North-East of England* HMSO London.

Monopolies and Mergers Commission (1997) *FirstBus plc and SB Holdings Ltd: A Report on the Merger Situation* The Stationery Office, London.

NERA (1997) *Evaluating Alternative Structures for the Bus Industry: A Report for the Confederation of Passenger Transport* NERA/CPT, London.

Office of Fair Trading (1989) *Highland Scottish Omnibuses Ltd: Local Bus Services in Inverness* OFT, London.

Office of Fair Trading (1992) *Southdown Motor Services: The Registration and Operation of Services 262 and 242 in Bognor Regis* OFT, London.

Office of Fair Trading (1993) *Thamesway Ltd: The Operation of Local Bus Services Commencing in, Terminating in, or Passing Through Southend-on-Sea* OFT, London.

Office of Fair Trading (1997) *The Effectiveness of Undertakings in the Bus Industry* prepared for the Office of Fair Trading by NERA, OFT, London.

Oldale, A. (1998) *Local Bus Deregulation and Timetable Instability* LSE Economics of Industry Discussion Paper 21.

Phlips, L. (1995) *Competition Policy: A Game Theoretic Perspective* Cambridge University Press, Cambridge.

Shaked, A. and Sutton, J. (1982) Relaxing Price Competition Through Product Differentiation *Review of Economic Studies*, 49, 3-13.

White, P. and Tough, S. (1995) Alternative Tendering Systems and Deregulation in Britain *Journal of Transport Economics and Policy*, 29, 275-289.

White, P. and Farrington, J. (1998) Bus and Coach Deregulation and Privatisation in Great Britain, with Particular Reference to Scotland *Journal of Transport Geography*, 6, 135-141.

Notes

1 In a December 1997 report for the Confederation of Passenger Transport, NERA argued that expansion of the quality partnership approach would be a more effective way of improving the bus industry than would re-regulation in the form of either route franchising or area.

2 The Magicbus operation is a brand name of Stagecoach Manchester, with buses operating from the same depots as other Stagecoach services. It runs buses 24 hours a day along the Wilmslow Road corridor, reputedly the busiest route in Europe, and one which is dominated by the student market. While buses run under the Stagecoach Manchester brand name start in various distant suburbs, the Magic buses only run from where the student areas begin, and are overwhelmingly filled by students, especially in the early hours. Magic buses are older vehicles, but only as old as those on many routes in the city. They compete with a large number of other operators on the routes, including some cut-price ones. The frequency on the corridor is 120 per hour, so that there is

often a choice of bus at a stop. Some people other than students have been observed waiting for a Magic bus, though one reason may be that a Stagecoach weekly ticket (which is valid on all Stagecoach buses) is £1 cheaper if purchased on a Magic bus. (Thanks to Dr David Forrest, of Salford University Economics Department, for this information.)

3 NERA have reported on the effectiveness of a selection of behavioural remedies, and one structural remedy (in Portsmouth), imposed in competition policy cases in the bus industry (OFT, 1997).

4 Apologies to the Scots and Welsh for this section of the paper. See White and Farrington (1998) for a recent review of the situation in Scotland.

5 Kennedy (1965) shows bus tendering in London reduced costs by twenty per cent.

8 How Effective is Competition? - A Comment

CHRIS NASH

Introduction

The papers of Dodgson and Preston offer a relatively rare opportunity to compare and contrast the experience of two privatised modes of public transport in Great Britain - bus and rail. Two modes - two forms of competition. In the bus industry, with complete freedom of entry (except for quality licensing) competition is mainly between different companies actually offering services in the market, although there is also significant competition for the market in the form of competitive tendering for subsidised services (and all services in London). In the rail industry competition is mainly for the market in the form of bidding for franchises, although there are some cases where more than one company operates between a particular pair of points, and the Regulator is considering gradually liberalising entry to make this possible in an increased number of cases. There is also free entry into rail freight, although actual competition there is very limited with essentially only three operators in Great Britain, one of which has a very dominant position. This comment will consider the bus and rail industries in turn before seeking to reach conclusions.

The Bus Industry

In the case of the bus industry, Dodgson considers a key issue to be the degree to which the industry is contestable, and few would argue with that. There is a question though as to whether even perfect contestability would be sufficient for markets to deliver socially efficient bus services priced at marginal social cost. There are two main cases for subsidy of bus services. The first is that their marginal social cost is significantly below average cost. This occurs despite considerable evidence that economies of scale in bus company size are very limited, and it occurs because of an effect known commonly as the Mohring effect (Mohring, 1972). When extra bus passengers occur, the operators have two choices. Either they accommodate the extra passengers in existing vehicles, at low marginal cost (even if larger vehicles are needed there are substantial economies of vehicle size). Or they put on extra services, giving a more frequent service to existing passengers. Either way, the average social cost, including the time of the passengers, is reduced.

Politically there is much more support for an alternative case for bus subsidies; the second best one of relieving congestion by diverting passengers from the under-priced car mode. Evidence (e.g. Glaister, 1987) suggests that of the two motives quantitatively the Mohring effect is by far the more important. But both give justification to a policy of subsidising even (perhaps especially) those services which are able to break even. Now it is very difficult to devise a way of giving these subsidies that gives appropriate incentives to competing operators to adjust fares and services towards the optimum (Else, 1985). It is far easier to do this by competitive tendering for services planned by a public authority (though of course this opens up the argument of Government failure - that the authority in question may fail to optimise due to a lack of appropriate incentives on it).

Suppose however that subsidies are not available for the majority of services. Would free competition lead to a second best optimum in terms of fares and services? The answer Dodgson gives is no. The bus market is far from perfectly contestable. As a result, the 'curious old practices' which Foster and Golay (1986) declared would not occur with deregulation are back in force. Schedule matching, headrunning, hanging back, predatory pricing and service levels are all daily occurrences in the British bus industry. Whereas Foster and Golay essentially argued that firms would accept the inevitability of new entry and adjust to make the best of the situation, what we actually see is a fight to the death in the majority of cases of competitive entry. In a truly contestable industry this simply would not make sense.

Yet it is clear that the threat of new entry is a major force in getting costs down. It has been argued that competitive tendering is a powerful way of exploiting this threat. Competitive tendering leads to a much more contestable market, since once the sealed bids are opened and the decision taken, the loser has no opportunity to retaliate for several years (other than by operating the service without subsidy in competition with the winner, which arguably is a clearly predatory action that should be outlawed).

Dodgson refers to a study by NERA which prefers the 'quality partnership' approach to improving bus services rather than universal competitive tendering. He argues that the evidence that competitive tendering works much better than competition drawn on a comparison between London and the rest of Great Britain is spurious, because demographic and other trends are much more favourable to the bus in London than elsewhere. But it is not clear that quality partnerships could function in the absence of either a dominant operator or entry controls to ensure that new entrants meet the required standards of vehicles and services - otherwise they would be susceptible to competition from cut price operators who did not meet the required standards. If such controls are introduced 'quality partnerships' themselves may make bus services less contestable, by confining entry to high quality operators. Moreover, it remains unclear whether it would be possible in this way to restore appropriate combinations of fares and services without regulation of at least one of these variables.

Dodgson finally turns to the role of the competition authorities, which will be discussed more fully in a later paper. If the above argument is accepted then this is a very difficult role. It appears that in the bus industry, the threat of competition is generally more desirable than competition itself. Thus preserving an industry structure that makes this threat real is very important. But on the road competition needs heavy regulation, and the freedom of operators to set their own fares and services and the requirement that they do not collaborate over these decisions may actually be very damaging in terms of the outcome to the user.

The Rail Industry

Turning now to the rail industry, we find a much more complicated position. The rail industry is subject to substantial economies of density in terms of infrastructure use as well as the Mohring effect (arguably the bus industry is too but it usually does not have to pay for its infrastructure). Thus the arguments for subsidy, and for intervention to influence the pattern of services and fares that commercial operators favour, is even stronger than for bus. Despite the fact that a new entrant requires relatively little capital expenditure (it can lease rolling stock and buy access to the infrastructure, including maintenance facilities), barriers to entry in the form of the necessary safety case, trained staff with the appropriate route knowledge, time consuming negotiations and the likely length of leases on rolling stock, remain strong. These barriers are of course most easily overcome by a company that is already an operator on another route. Also the speed of retaliation is much less than in the bus industry. Whilst the requirement to give 42 days' notice has been argued by some to be a significant problem for contestability in the bus industry, in the rail industry the terms of the franchise agreement and agreements over dates for revising fares and services may impose a much longer delay in retaliation, and indeed it may well be totally impossible to negotiate the set of access rights that either the entrant or the incumbent would desire. Thus schedule matching and racing between stops that are so often seen in the bus industry are simply not possible to the same extent in the rail industry because of the constraints of the infrastructure. But the risk of getting locked into a position where competition involves wasteful duplication of services a few minutes apart from each other is just as great if not greater in those cases where competitive entry does exist.

Preston points out that by the time of refranchising, if they meet their business plans, most of the existing InterCity and London and South East rail operators will be profitable. This means that at refranchising it would be possible to go for a 'bus industry' approach whereby operators were invited to bid for slots to operate what services they wished commercially, and competitive tendering was confined to the rest of the system. But would this be a desirable approach? The discussion suggests that this would be even worse than for the bus industry, with competition both more constrained and more damaging when it did occur. This lends support to the approach currently being followed by the Office of the Rail Regulator, which looks for a gradual easing of the conditions of entry, but with each case being judged on its merits and with entry only being allowed where it offered beneficial innovation in terms of fares or services, rather than simply duplicating existing services.

So it appears that competition for the market will remain the main force in the rail sector rather than competition in the market. Does competition for the market work for rail? The answer Preston gives seems to be a qualified yes. Certainly the bidding for the first round of franchises suggests that this form of competition can achieve substantial reductions in costs and increases in traffic and revenue over time, unless many bidders were hopelessly optimistic in their bids. But worries still remain. Are operators currently being encouraged to flood the market with excessive service levels to keep out competitors, rather than keeping fares down? (a tactic certainly common in the bus industry, but even more effective in rail, in that once the track capacity is filled up it typically takes lengthy and costly investment before further services are possible, and neither Railtrack nor an entrant would be likely to finance this for the sake of a speculative competitive entry). Are the penalties for failing to perform in terms of punctuality, reliability, cleanliness and overcrowding strict enough, and should such penalties be more widespread in the parts of the system where currently competition with other modes is seen as sufficient to enforce quality standards? Will sufficient investment take place, particularly in the latter part of a franchise, when even investment in infrastructure or rolling stock funded by third parties may be unwelcome in that it detracts from current performance for the benefit of future services? Are long franchises or franchise extensions a partial answer to this problem, or will they give too much monopoly power to existing operators? And with the trend towards concentration of the bus industry, and the fact that the leading bus operators also hold or are a member of the consortium holding several rail franchises each, will the bidding for future franchises be as competitive as it was the first time round? There is plenty of work ahead for the new Strategic Rail Authority and for the Rail Regulator in resolving these issues.

Conclusion

So the big debate between proponents of competition for the market and competition in the market goes on, and has an even more complicated terrain on which to carry on the fight in the rail industry than it did in buses. But it is also clear that the simple characterisation of the choices in these terms is far too simplistic. Whichever broad approach is adopted, many factors concerning the way in which it is implemented are crucially important. To what extent are fares and the level and quality of services controlled? What are the conditions which must be satisfied by new entrants before they are allowed to come into the market? What is the role of the regulation and competition authorities? What behaviour should they allow or disallow? The emergence of the idea of 'Quality partnerships' as the way forward in particular begins to open up the middle ground between the completely laissez-faire approach to fares and services of total deregulation and a situation where fares and services are totally controlled by public bodies. Clearly this is an area for fruitful research in the coming years.

References

Else, P.K. (1985) Optimal Pricing and Subsidy for Scheduled Transport Services. *Journal of Transport Economics and Policy,* 19,3, 263-80.
Foster, C.D. and Golay, J. (1986) Some Curious Old Practices and their Relevance to Equilibrium in the Bus Industry. *Journal of Transport Economics and Policy,* 20, 2, 191-216.

Glaister, S. (1987) *Transport Subsidy.* Policy Journals, Newbury.

Mohring, H. (1972) Optimisation and Scale Economies in Urban Bus Transportation. *American Economic Review,* 62, 591-604.

9 Regulation of Railways and Roads: Is the Current Framework Adequate?

BILL BRADSHAW

Introduction

The purpose of regulatory institutions is to set out the detailed rules within which competition takes place. These include safety and quality standards and, where it is considered that the market provides inadequate discipline, regular reviews of the permitted level and structure of charges. A further role of institutions is to create and maintain conditions in which the cost of capital is held at levels which will allow investment to take place, and to require such investment as is needed to meet strategic, economic, environmental and safety objectives. These institutions work within the law, and each application or use of regulatory discretion and intervention is judged on its own merits.

The level and structure of charges for using the railway have been regulated ever since the nineteenth century, when railways enjoyed a near-monopoly in many markets. The remnants of this regulation survived on the statute book until recent years and continued to be effected through a process of informal ministerial intervention during most of the period of public ownership.

This paper reviews the effectiveness of the institutions currently responsible for the regulation of the rail and road transport industries in Great Britain with regard to investment, service quality, fares, safety and competition. Some suggestions for reform are offered.

Regulation of Investment

Railways

At privatisation, it became the responsibility of Railtrack to maintain, renew and develop the network and to set out its plans in an annual Network Management Statement (NMS). However, dissatisfaction with the first two of these documents, and concern that insufficient investment was being made in the network, led the Regulator to secure an amendment to Railtrack's licence in September 1997. This modification strengthens the Regulator's powers to enforce the implementation of Railtrack's NMS. In particular, it established a new general duty on Railtrack in respect of the maintenance, renewal and development of the rail network, and a series of further detailed and enforceable obligations in support of this duty. The amendment also gave the Regulator powers of investigation and enforcement if Railtrack fails to satisfy its obligations without good reason. Finally, obligations were placed on Railtrack both to consult with train operators and funders and to meet their reasonable expectations to the greatest extent reasonably practicable.

When Railtrack's third NMS was published in March 1998, the Regulator commented that it provided passengers and freight customers with greater detail about Railtrack's plans, but that the statement, as it stood, contained very few firm commitments to deliver significant improvements across the railway network. For example, the NMS identifies potential bottlenecks in the network, where capacity approaches 90%, but most of these are subject to 'study and evaluation' with very little commitment to actual investment. The only exceptions are the West Coast Main Line schemes and Thameslink 2000. The long time-scales involved, with schemes possibly coming on stream between 2001-2007 (mostly around 2005) indicate that growth in demand for rail transport, requiring physical alteration of the infrastructure, as opposed to running longer trains or re-timing, will not be available in time to absorb early expansion.

Except on journeys over the West Coast Main Line, the improvements in journey times projected are mostly very modest. Railtrack has also taken a cautious view of the likely increase in rail passenger demand, with a central forecast of growth of 15% over ten years within a range of 5% to 30%. The higher levels were associated with the introduction of road pricing, workplace parking taxes and steadily increasing fuel prices. Viewed historically, these forecasts may be justified, but they do not accord with the wish of the Environment, Transport and Regional Affairs Committee to see the Strategic Rail Authority developing ambitious targets for growth (HC 286-I, 1998)[1]. Neither do they accord with the growth targets of some Train Operating Companies. When allowance is made for the strong growth targets on the upgraded West Coast Main Line, Railtrack's targets for the rest of the network appear to be very modest indeed. With regard to the freight market, Railtrack's base forecast is for growth of 80% within a range of a 40–200% increase - the top end of this range coinciding with the declared targets of English Welsh & Scottish Railway. . In the case of Railtrack, this forecast is conditional upon the restriction in the general use of 44 tonne lorries, road-pricing and trucks paying their full environmental costs.

Maintenance expenditure is set to fall between 1997/98 and 2007/08, from £740m per annum to £490m per annum, with cost reduction contributing to a higher quantum of work accomplished. Renewals expenditure falls over the same period from £1 billion per annum to £610m per annum. Enhancements to the network over the next ten years will cost £3.2 billion, including track (£540m), signalling (£815m), structures (£250m), electrification (£280m), and stations (£265m).

Addressing the issue of financing the work, Railtrack displays reluctance to undertake enhancements, unless there are either additional access charges and guarantees about future use from the Director of Passenger Rail Franchising, or extra subsidies from the passenger transport executives (PTEs), local authorities etc. There is clearly an issue to be resolved here about the treatment of capital expenditure by the Regulator at the forthcoming review of track access charges. With the long time-scales of enhancement work being quoted by Railtrack, virtually none of the existing franchisees can commit themselves to additional access charges. It is notable that almost all the developments and enhancements listed in the NMS were initiated by British Rail and, although Railtrack has existed as an independent organisation for four years, there has been little innovation by the company which is near to being implemented. With lead times of seven to ten years to plan and commission even modest track reconstruction schemes, there is an obvious discontinuity between franchise lengths and Railtrack's capability to deliver significant enhancements which have been developed in partnership with the train operating companies.

The third NMS shows Railtrack to be extremely risk-averse and unwilling to commit itself to network enhancement and development without guarantees about future access charges. The West Coast Main Line agreement with Virgin, where there is a 15-year franchise, is an obvious exception. But how much risk is Railtrack really facing? By 2003, almost all the franchises involved in providing InterCity or London commuting services are likely to be re-let at a premium, with service levels at least at today's levels and in all probability running more trains. Local services in PTE areas are likely to be maintained as these authorities have shown great loyalty to railways. Even a pessimistic analysis of the future would define a large part of the network where the continued payment of access charges could be virtually guaranteed. The level of these payments will depend upon the forthcoming regulatory review and the capability of the network to deliver a more reliable, frequent and faster service. These features are what the Regulator and the SRA will be looking for and which will enable future train operators to pay more. Railtrack appears to be unwilling to interpret the role of 'steward of the network' as taking a lead in developing the railway to absorb the growth in railway traffic which Government apparently requires. Railtrack's unwillingness to show this leadership makes it inevitable that a SRA will take over the planning role and that much more intrusive intervention by the Regulator can be expected to ensure Railtrack provides all the information needed for planning.

Following a consultation into whether the third NMS had identified and actively sought to rectify the needs of train operators and funders of the railway, the Regulator secured improvements in four areas.

- Performance - the consultation process showed that train operators and funders believed that there was greater scope for performance improvements than was detailed in the NMS. As a result, Railtrack has agreed to reduce by 7.5% the delays that it causes in 1998/99, and to continue its analysis of train delays. In January 1999, Railtrack will publish new targets for improving performance for the two years to April 2001.

- Track quality - Railtrack will draw up a programme aimed at eliminating poor quality track on the network by April 2001.

- Network capacity - Railtrack has also committed itself to completing evaluations of options to address capacity problems by November 30th 1998.

- Meeting the needs of train operators and funders - Railtrack has agreed to specific improved, timetabled arrangements for establishing the requirements of the train operators and funders. It has said that it will also improve the transparency of procedures for initiating, evaluating and carrying out enhancement investment.

Railtrack is also committed to a major consultation programme, aimed at determining the reasonable requirements of train operators, PTEs and OPRAF.

At the time that BR was privatised, critics argued that the sharing of tracks between operators, some of which will be competing with each other for business and for opportunities to run trains, would make for difficulties in planning and securing investment. They also argued that there were problems with letting short franchises, as it would not be worthwhile for franchisees to invest in new rolling stock that has a very long lifetime. Indeed, the considerable sunk costs involved in railway operation contrast with the much more flexible investment decisions faced by those involved in road and air transport.

Many of the problems faced by the railways are manifest in the case of the West Coast Main Line. The Regulator has approved a revenue sharing arrangement between Virgin Trains and Railtrack to upgrade the West Coast Main Line, the aim of which is to spread the risk of delays to the work programme and incentivise Railtrack to ensure the network improvements are undertaken as smoothly as possible. The deal has not only involved a much longer franchise than was envisaged when privatisation was planned, but has allowed a partial re-integration of the interests of infrastructure provider and principal train operator. Other operators who use the track have expressed concern that the arrangement could lead to discriminatory treatment of train operating companies by Railtrack. The agreement has also raised concern regarding what Government and the regulatory institutions would do if the contractual parties came back and said they were unable to deliver the improvements in the time allowed. If there is a case for rescuing the Channel Tunnel Rail Link then the West Coast Main Line would surely follow?

The key problem facing the regulatory institutions is how to assess whether proposed investments represent good value for money and whether the order of priorities fits in with the larger, strategic, picture. It is very difficult to see how satisfaction can be obtained in this matter unless one moves to the extreme that the Regulator, or perhaps the Strategic Rail Authority, actually commissions or sanctions investment. Railtrack can obtain competitive quotations for some work, but it is by no means clear that this will be sufficient to satisfy critics that best value is being obtained.

In April 1999, Mr Chris Bolt, acting Rail Regulator, published a report prepared by consultants Booz Allen & Hamilton (1998)[2], that details Railtrack's performance as owner and operator of the rail network over the past four years, as well as considering its prospective performance to 2001. The report represents an integral part of the periodic review of Railtrack's access charges. Booz Allen found that the Rail Regulator could reasonably have expected greater improvement in performance at a national level, and that generally, Railtrack's physical activity in renewing assets has been below expectations. It appears that Railtrack has focused its investment efforts on assets that are likely to generate performance improvements in the short run, such as the renewal of rail, rather than investments in long-term drivers of performance and quality. The structure of incentives that faces Railtrack has led it to be reactive to schemes proposed by other parties, rather than entrepreneurial. As a result, there has been little increase in capability. The overall level of expenditure on enhancement has been relatively low, at 14% of the total asset management plan, renewals and enhancement expenditure. On the plus side, Railtrack has delivered significant improvements in operational and management processes, and has exceeded prior expectations in terms of productivity improvement and cost efficiency. Railtrack has also been proactive in developing new contracting forms for the rail industry. However, there are several areas in which Railtrack could have delivered additional efficiencies.

Mr Bolt is now seeking the views of the train operators, funders of the railway, Railtrack and others on the report's conclusions. In addition, Mr Bolt is seeking views on whether or not the plans set out in Railtrack's Network Management Statement, published on March 25th 1999, meet the reasonable requirements of train operators and funders, as required under Railtrack's network licence. Booz Allen & Hamilton is currently assessing Railtrack's future expenditure needs and the scope for efficiency savings after 2001.

At the National Rail Summit in February 1999, Dr John Reid, Minister of Transport, announced that the Government is prepared to re-negotiate existing franchises in return for guarantees of better services for passengers. Within the next six years, 17 of 25 franchises are up for renewal. Dr Reid said there would be strict criteria against which the re-negotiations would be judged and the delivery of substantial passenger benefits would be a key consideration in deciding whether or not to open discussions with individual train operators. The six key criteria are as follows:

- current performance;
- proposed accelerated or new investment;
- a commitment to more demanding enforcement regimes and tougher Passengers' Charter targets;
- a commitment to promote integrated transport;
- a willingness to give passengers a greater say in standards of service; and
- value for the taxpayer.

Dr Reid warned that there would be no automatic right to sit at the negotiating table, and the Government plans to issue new objectives, instructions and guidance to OPRAF on how to take the process forward. Dr Reid said that only a limited number of franchises will be renegotiated, and operators will have to compete to be part of the first tranche.

Franchise re-negotiation could involve three areas:

- changes to contractual outputs within existing franchise terms;

- securing extra investment or other benefits by underwriting future revenues; or
- extending existing franchises.

However, it is still unclear whether the renegotiation process will deliver much in the way of benefits to users unless franchisees are given what will amount to security of tenure.

We seem locked into a system where Railtrack and the franchisees decide what investment takes place, and the regulatory institutions are in a relatively weak bargaining position. The regulators have little useful information and very few benchmarks or other comparators of efficiency. It is not clear that the Strategic Rail Authority will be much better placed unless it has money to spend as a means of leverage on Railtrack in particular. However, it appears that the SRA will only have some £150m to spend over three years with which to eliminate bottlenecks, support local authority projects such as station re-openings and promote freight expansion (Financial Times, 25th May 1999)[3].

Roads

On current projections and without policy changes, traffic could grow by more than a third over the next 20 years and by more than half on trunk roads. This would result in serious congestion on a quarter of the trunk road network by 2016. Investment in new roads must come from Government, either directly or through a PFI mechanism. The difficulties of securing planning consent for new highways means that only Government can reasonably shoulder planning risks. In the first of the daughter documents to the Integrated Transport White Paper, DETR (1998)[4], which reported the outcome of the Roads Review, the Government did not use these forecasts to justify an acceleration of road building. Instead, reference is made to the SACTRA report, which is generally accepted to have signalled the end of the 'predict and provide' approach by showing that new road building simply generates more traffic, SACTRA (1994)[5]. The new approach taken by the Government to trunk road investment is outlined as follows:

- improve trunk road maintenance;
- make better use of existing roads by investing in network control and traffic management measures and in safety improvements; and
- tackle some of the most serious immediate problems through a carefully targeted programme of improvement schemes.

It is proposed that the trunk road network be managed as part of a series of *transport networks* - including roads, railways, inland waterways, airports, ports - which have good connections between them. The Government inherited a £6 billion road building programme of around 150 schemes, scaled down from over 500 in 1990. Decisions on 14 schemes were made in the accelerated review in July 1998. The remaining schemes were assessed using a 'radical approach' to building and operating roads. This involved the replacement of traditional cost-benefit analysis (CBA) with a method which assesses road schemes against a number of qualitative criteria, including environmental impact, safety, accessibility, economy and integration. Several schemes with high benefit-cost ratios have been dropped, while others that ranked much lower on the basis of a strictly economic appraisal are to go ahead. The result is a new 'targeted programme of improvements', which comes down to 37 schemes totalling £1.4 billion, all of which will be started in the next seven years.

The Government has clearly rejected the option of significantly increasing the capacity of the road network on environmental, social and economic grounds. If the impacts of gridlock on the roads are to be reduced, it is therefore vital that the capability of the rail network be strategically reviewed so that:

• more trains can run;
• trains run reliably and punctually;
• connections are always made;
• stations are safe and secure and provide good interchange.

To reduce the upward trend in traffic growth by about one-third over the next ten years, rail use would have to be doubled and bus use would have to increase by 50%. That is a prodigious task, but vital if the targets for greenhouse gas reduction agreed at Kyoto in December 1997 are to be achieved.

Quality of Service Regulation

Railways

In the debate leading up to privatisation, it was contended that detailed regulation of punctuality, cleanliness and other aspects of customer service would be unnecessary. This was because, or so the argument ran, 'an industry keen to win business and focusing on customers would, unlike British Rail, run to time and keep trains, stations and lavatories sparkling'.

However, it was realised that it would be necessary to regulate the provision of seating capacity because, with some operators having peak vehicle requirements greatly exceeding those at the off-peak, the financial advantages of under-providing rolling stock would be very attractive.

In the event, the quality of the railways has, with some notable exceptions, been very disappointing. The regulation of quality is currently divided between the Rail Regulator and the Franchise Director. The report issued in March 1998 by the transport sub-committee of the Environment, Transport and Regional Affairs Select Committee found that the current system of dual regulation has resulted in obscure boundaries and poor supervision of quality standards. When launching the Transport White paper in July 1998, Mr Prescott stated that the railways have suffered from a lack of strategic planning, fragmentation of services and profits being given priority over passengers. The solution advanced is to create a Strategic Rail Authority. It is envisaged that this will take over the functions of the Office of Passenger Rail Franchising (OPRAF) and its responsibilities will include:

- managing the passenger franchises;
- administering subsidies;
- allocating freight grants;
- ensuring that the railway runs as a single network, integrated with other modes;
- protecting the consumer;
- sponsoring the Central Rail Users' Consultative Committee and Rail Users Consultative Committee passenger groups; and
- administering two new funds totalling £100m (the Infrastructure Investment Fund will support projects to address capacity constraints at 'pinch-points' on the network; the Rail Passenger Partnership Scheme is targeted at innovative proposals to encourage integration between modes, and modal shift to rail).

The SRA is also to ensure operators' structure and market fares to offer 'value for money', and 'to reflect the fact that the railway... needs to be marketed as a network to encourage a switch from car to train'.

In addition, the Railways Act is to be amended to enable imposition of penalties in respect of past breaches that have ceased and more rapid enforcement action. The Rail Regulator is to be given a new duty to take account of the Government's broad policy for the passenger and freight railway, and new duties relating to integrated transport and sustainable development.

Many of the responsibilities of the SRA, as outlined above, are clearly not strategic in nature. The danger is that by taking on many day-to-day tasks such as monitoring operational performance and sponsoring consumer representation, the SRA will not be able to concentrate on the strategic role of ensuring that the network is capable of absorbing more traffic.

An interesting feature of the management of punctuality on the railway is the performance regime (Schedule 8 of the Track Access Agreement). Under this mechanism, Railtrack and train operators compensate one another for delays. Railtrack measures and attributes responsibility for delays, which have to be agreed with the train operator. In cases where Railtrack cannot allocate responsibility, half of the incidents are allocated to Railtrack and half are split between Railtrack and the train operator in the ratio of responsibility for allocated delays. Railtrack must pay compensation if average lateness over a four-week period is higher than the agreed limit for a given route group. TOCs pay a supplemental charge to Railtrack to cover the estimated net cost of operating the performance regime. Making payments for delays inflicted on other users is unique to the railways. If we ever invent 'Roadtrack', it will be interesting to see whether road maintenance contractors or the owners of lorries that shed loads in the morning rush hour are obliged to compensate other road users for the delays they cause.

There is great concern about quality of railway services. The annual report of the Central Rail Users' Consultative Committee (CRUCC) recorded an astonishing rise of 103% to 19,972 in the number of complaints received by the eight RUCCs during the year ended March 31st 1998, compared with 1996-97. This is the highest number of complaints ever recorded and follows a year in which complaints actually fell, following a climb from 8,000 to 11,600 during the years leading up to and during privatisation. Compared with 1988-89, when British Rail carried 1% more passenger-miles than the 25 franchised TOCs achieved in 1997-98, complaints have quadrupled. The CRUCC report notes that such regional increases in complaints as there were in 1996-97 were confined to the south and east. However, 'during 1997-98 complaints shot up in every part of the country' with the biggest increases in Scotland (120%) and western England (200%).

Combined with deteriorating reliability and punctuality figures published by OPRAF for the year to June 27th 1998, the volume of complaints prompted Mr David Bertram, Chairman of CRUCC, to comment that 'the realities of privatisation are proving to be too often a painful experience for passengers.' While welcoming many improvements in services that the TOCs have introduced, and the recent upsurge in traffic, Mr Bertram notes that 'in tandem with this growth has come the inevitable problem of overcrowding, with OPRAF already reporting half the TOCs serving London above the limits set.' He observed that overcrowding 'is now spreading rapidly to longer-distance services and increasingly on routes into other major conurbations too.' Bertram believes the 'realities' of the 1993 Railways Act 'are increasingly evident', and describes 'inadequacies' in protecting the interests of passengers as 'worrying'. He warns that the Act 'does not encourage growth' so 'the recent high level of investment could dry up as franchisees' pay-back time-scales shorten.'

There is a fundamental question as to whether a railway - which is the quintessential integrated production process - can ever really respond to the passenger in quality terms when so many of the elements of production have to be co-ordinated on a minute-to-minute basis, in the absence of a command structure. The only way to avoid complaints is to sink to a sub-optimal system of operation where the assets are under-worked to leave margins during which any shortcomings can be recovered. Basically this means timing fewer trains, more slowly, using more rolling stock than necessary and being thoroughly unambitious in the use of capital and personnel assets. It is neither the way to run a business or a railway. I see quality regulation in the railway developing into a form of arid trench warfare - a 'what can we get away with?' culture rather than one of 'how can we serve the passenger?'

Roads

Quality regulation in road freight transport means only that legal safety standards are observed. Otherwise the market will pay for the quality it wants. Buses and coaches similarly will need to conform to decent standards if passenger growth is to be achieved. The Transport Select Committee Report on the consequences of ten years of deregulation in the bus industry found that the required quality standards were very low and were inadequately enforced, largely due to a lack of resources. The inquiry also found that the licensing authorities were finding it extremely difficult to monitor compliance with requirements such as registration of services or to take effective remedial action if breaches were detected (HC 54-I, 1995)[6].

Traditionally bus and coach users have had nobody to appeal to if they believe they have had a raw deal and the operator will not listen. As Chairman of the newly created Bus Appeals Body I hope that we can ensure that operators have proper procedures in place to receive, investigate and settle complaints in a timely way. Where the complainant is not satisfied, the case will be reviewed by representatives of the industry and the users as well as by myself. We will make determinations and report our findings to the Traffic Commissioners. At the end of 1999 we will distill our findings into what I hope will be some lessons for the industry and guidance in good practice. I hope that working with the grain we can achieve speedy and satisfactory outcomes. This approach is worth a try and I shall look forward to reporting on the experience. As an institution we are young (not personally) and very cheap.

Fares Regulation on the Railways

Regulation of public transport fares is relatively light in Britain. What is clear is that fares on public transport have risen over the past twenty years or so much faster than the costs of motoring or RPI in general (see Table 1 below).

Table 1   Retail prices index: transport components: 1986-1996

Year	All items	Petrol and oil	All motoring	Bus and coach fares	Rail fares
1986	84.9	93.5	85.7	81.9	81.5
1987	88.4	94.3	90.7	86.7	85.7
1988	92.8	93.3	94.8	92.7	91.6
1989	100.0	100.0	100.0	100.0	100.0
1990	109.5	111.8	106.1	105.5	108.8
1991	115.9	120.1	114.0	120.3	120.1
1992	120.2	123.5	121.7	128.8	128.9
1993	122.1	133.3	126.9	134.4	137.9
1994	125.1	139.5	131.3	138.0	144.1
1995	129.4	146.6	133.7	143.0	150.4
1996	132.5	154.1	137.7	148.4	156.0

Source: DETR (1997), Transport Statistics Great Britain.

Some local authorities still specify fares on local rail services, but bus fares are not regulated and only a relatively small proportion of railway fares are subject to a price cap specified in the franchise agreements. The fares which are constrained include unrestricted standard class return fares and certain single fares for short distance journeys, 'Saver' fares (which do not include 'SuperSaver' fares) on other journeys and certain standard class season ticket fares including all those for weekly season tickets. For flows subject to the compulsory inter-availability requirement, only inter-available prices are controlled and the lead operator is obliged to create fares of the specified type(s). If there is no compulsory inter-availability for any flow, and therefore no lead operator, the Franchising Director imposes equivalent obligations on one or more of any TOCs which provide direct train services on the flow in question. For journeys of over about 50 miles, except those wholly within the south-east, BR used to offer a saver fare. For each flow where a saver fare existed as at June 1995, the Franchising Director requires a return fare with the following characteristics to be available at a price which is capped by reference to the price of the saver fare as at June 1995:

- valid for at least one month;
- valid on any day of the week;
- valid at any time of day, except that it need not be valid before 1030 (Mondays-Fridays) or for journeys leaving Greater London and certain major stations near London between 1500-1900 (Mondays-Fridays).

For other journeys, the Franchising Director requires a return fare which is valid at all times of the day to be available, at a price capped by reference to the price of the unrestricted open (or one-day) return fare as at June 1995. For those journeys where BR offered a standard class seven-day season ticket as at June 1995, the Franchising Director requires an equivalent fare to be available at a price capped by reference to the price of the seven day season ticket as at June 1995. This applies to both the adult and child prices. Fares regulation extends to through journeys, which may involve interchange between London stations by Underground and to London Travelcard season tickets, but not to other journeys involving the use of services other than those of franchise operators. It also excludes fares for journeys where a PTE is entitled to set prices.

For the three years from 1 January 1996, increases in capped fares are not permitted to be more than the RPI from the 1995 base price. For the four years from 1 January 1999, the price cap for such fares is RPI minus one. Unless the Franchising Director (probably the SRA by then) decides otherwise, the price cap from 1 January 2003 will continue to be RPI minus one. There is some limited scope for individual fares to exceed the cap where these are balanced by other controlled fares being held below cap levels.

In the London, Cardiff and Edinburgh commuter markets, an extended approach is taken to regulating the price of fares. The Franchising Director has the power to adopt a similar approach in markets where he considers rail to have particular market power. Regulation applies to an extended range of fares including all standard class season tickets and, within defined areas, unrestricted single and return standard class fares. Price control is exercised by reference to tariff baskets containing all relevant fares, weighted broadly according to the income which a franchise operator derives from each. The overall weighted average of the prices of fares in tariff baskets is capped.

To ensure that passengers are both compensated for poor performance and pay for improved performance through fares, for operators subject to the OPRAF performance regime, the caps on the price of fares in the London areas' tariff baskets are generally adjusted to reflect the quality of the franchise operator's performance. Quality of performance is determined under the OPRAF performance regime. Improvements or deterioration in performance may lead under this regime to the cap being adjusted by up to 2% up or down. The first adjustments applied in 1997 on the basis of a benchmark that is normally set at the higher of current performance or Passengers' Charter standards. Franchise payments are adjusted with the intention that the franchise operator is not permitted to benefit from both a fare increase and OPRAF incentive payments for the same quality improvement. The converse applies if performance deteriorates, and operators may be fined by OPRAF and obliged to offer fares rebates/season ticket discounts. Within tariff baskets, the prices of individual fares may increase above their June 1995 level by up to 2% per annum above the increase in the RPI. A franchise operator wishing to invest to improve service quality is able to request additional increases in regulated prices, which the Franchising Director considers. Fares for children under 16 (whether the price is otherwise regulated or not) must be available on terms which are no less favourable than would apply to the holder of a Young Person's Railcard.

Barry Doe reported (Modern Railways, July 1997) that Virgin increased its fares on Anglo-Scottish CrossCountry SuperSavers by 15%. This rise had made some of them dearer than the respective Saver. He reported that many of those SuperSavers exceeded the via London fare and were not supposed to be sold (Modern Railways, October 1997). Following this, the Regulator insisted that Virgin should do a printout of every fare sold and found that these fares had indeed been issued. In addition, even the winter Fares Manuals retained these incorrect non-London fares.

Barry Doe (Modern Railways, March 1998) has highlighted the fact that fares-capping is often misunderstood. Where held to inflation, it does not imply fares have to keep within any 12-month figure. Capping has a base level of June 1995, and between then and the January 1998 fares rise, the Prices Index rose 9.27%. A capped fare cannot currently be more than 1.0927 times what it was then. If inflation were 3% per annum, prices would rise 34% over a 10-year franchise as they compound in the way mortgage interest does. If a TOC held capped fares at 1% per year for nine years he would then be entitled to a 'final fling' of 23% in year 10 to catch up! Doe suggested that those franchises deciding not to re-tender might well be tempted to do just this.

Table 2 examines the 17 domestic TOCs that converge on London, indicating the percentage rise between June 1995 and January 1998 for all types of walk-on fare from London to a basket of destinations for each. ScotRail now has its own fares for the sleepers but did not in 1995 so cannot be compared. Where a TOC changes its fares, in June or September rather than January, these have been re-indexed to make the comparisons valid.

Thameslink has clearly maintained a policy of keeping peak fares well down, but raising off-peak substantially whilst LTS has done the opposite. The largest increases were as follows:

- Midland Main Line - SuperSaver ticket price increased 16% (mitigated only by its below inflation singles);
- Connex South Central with well above inflation in the off-peak;
- Silverlink for raising standard singles 13% and other fares by inflation; and
- Virgin Trains with above-inflation singles coupled with 21% on SuperSavers.

However, SWT has kept all types of fares at, or a little below, inflation. Great Western has kept the capped saver well under what it could charge and has introduced the walk-on Saver First. Wales & West has kept below inflation on all fares. In summary, a quarter of all fares have risen below inflation, half by inflation, and a quarter above, with off-peak fares coming off relatively badly, except for the capped Savers.

Table 2   Walk on fares compared for period (June 1995-1998)
(Inflation = 9.27% over the period)

Company	First single	Std single	Super Saver	Saver	Away break/ Stay-away	Cheap Day	Seasons (all types and both classes)
Anglia Railways	11	12	-	8	-	-	9
Chiltern Railway	9	10	-	9	-	6	11
Connex South Central	9	9	-	-	12	12	10
Connex South Eastern	11	10	-	-	9	11	11
Gatwick Express	14	7	-	-	-	10	10
Gt. Eastern	11	11	-	-	10	10	10
GNER	13	12	11	9	-	-	9
Gt.Western Trains	10	11	10	7	-	-	10
LTS Rail	-	14	-	-	-	5	9
Midland Main Line	7	8	16	6	-	-	9
Silverlink Train Ser.	9	13	-	-	10	10	10
S W.Trains	9	10	-	-	8	10	8
Thames Trains	10	9	-	-	10	10	8
Thameslink Rail	7	7	-	-	10	15	7
Virgin Trains	9	13	21	9	-	-	10
Wales & West	-	9	8	7	-	9	9
WAGN Railway	8	8	-	-	10	9	9

Source: Modern Railways, March 1998

Safety Regulation

Railways

Railways have traditionally been the subject of regulation, particularly in the areas of safety and charging. Safety regulation has a long and honourable record in safeguarding the lives of users and workers but, under recent intense media pressure, has moved in the direction where the costs of safety measures have exceeded, in terms of accidents avoided, the criteria used to justify expenditure on road safety. This has led to an interesting debate about how much it is worth spending to save a life on the railway as opposed to elsewhere. An example of this is the issue of Mk I coach modification and replacement deadlines. A consultation on this issue commenced in May 1998 with publication of a draft order under the Railway Safety Regulations 1998 requiring withdrawal from January 1st 2003 of Mk I stock not retrofitted with over-ride protection and central door locking. The draft also requires selective fitting of the Train Protection & Warning System by 2004. In the case of slam doors, the economic appraisal of the Health and Safety Commission's (HSC) showed that the costs of central locking are more than three times the safety benefits, mainly in terms of cost per life saved. Only two passengers were killed falling from slam doors on moving trains in the year to March 31st 1998, the same as 1996-97. Rail Business Intelligence (August 20th, 1998) asked how the HSC would react if responses submitted to the Health and Safety Executive (HSE) confirmed that one or both proposed Mk I modifications were not 'reasonably practical' on the basis of agreed valuations of cost per life saved. The reported response to this by HSC Chairman Frank Davies was that 'the public has no tolerance of rail deaths, there is zero public tolerance of deaths on the railway'.

On previous occasions, HSC and HSE officials have suggested that further 'judgement' is necessary when using the already inflated threshold of £2.65m per fatality avoided in major accidents because of 'public intolerance' (read: 'extensive media coverage') of rail disasters. However, in the case of individual rail fatalities not involving train accidents, the £900,000 applied to road deaths has been accepted by the HSE. Arbitrarily quadrupling this figure in the case of slam-door deaths - which are frequently the result of reckless or irresponsible behaviour and do not normally attract undue media attention - sets a potentially costly precedent for the industry. HMRI's Chief Inspecting Officer Vic Coleman has said that 'we will have to put an analysis before the Commission in due course', and promised to 'reconsider the proposals if that should prove necessary.'

Safety regulation is the overall responsibility of the Health and Safety Executive who have a staff under a Chief Inspector, Officer of Railways. It is interesting to see how the number of staff employed by the Railways Division of the HSE has grown since the activity was taken over at the end of 1990 from the former HM Railway Inspectorate of the Department of Transport (see Table 3).

Table 3 Staff in post in HM Railway Inspectorate

Year	Staff in post
1.4.91	34
1.4.92	52
1.4.93	63
1.4.94	71
1.4.95	85
1.6.96	82
1.4.97	88

Source: Health and Safety Commission Annual Reports

When this rise in staff numbers is compared with the decline in the numbers of staff employed on the railways, the number of railway workshops and the amount of railway rolling stock, questions arise about the efficiency of the HSE itself.

There is, however, another dimension to the regulation of safety and the railways and that is the mechanism of the 'Safety Case'. Following the traumas experienced by the railways in the fire at King's Cross Underground Station (November 1987) and the crash at Clapham (December 1988) the management of safety underwent fundamental review and a much more formal system was introduced. In the privatised railway, safety is primarily the concern of Railtrack, which is responsible for validating the safety standards of TOCs, as well as operators of light maintenance depots and stations, under the overall supervision of the HSE. Each participant must present a 'Safety Case', which in its simplest terms, describes the systems which have been put in place to reduce the risk of an accident to a level which is as low as reasonably practical (ALARP). The system has been criticised for being bureaucratic, slow and costly. In a speech to the Institution of Mechanical Engineers on March 1st 1996, the

Franchising Director, Mr Roger Salmon, drew attention to five types of new trains that were in storage because they were waiting for safety clearance from Railtrack. This was while older rolling-stock, which would not meet today's higher safety standards, continued to be in use.

In the days before privatisation, BR stood as its own insurer and was able and willing to accept responsibility for accidents such as that which happened at Clapham. Railway accident inquiries have now moved to a situation where the media and the police are placing greater priority on finding somebody to blame than on finding out the cause of an accident, and making recommendations. Under the present situation, it is often the case that several contracting parties have an interest in avoiding blame, which is unsurprising considering that directors of companies and those immediately involved in an accident are faced with the very real threat of criminal charges. The whole process for inquiries has, therefore, become legalistic and drawn-out, as exemplified by the time it took to deal with the accident which took place at Watford on August 8th 1996. This accident was caused when an evening rush hour train from Euston bound for Milton Keynes collided with an empty Euston-bound train south of Watford Junction in Hertfordshire. One woman died and 60 passengers were injured. Soon after the accident occurred, Railtrack carried out a formal three-day (internal) enquiry, and passed its findings to the HSE. However, the criminal investigation led to a manslaughter charge being pressed against the driver in January 1997, and it took over a year for him to be acquitted (March 1998). After the criminal proceedings had closed, the HSE was finally free to publish its report into the incident in April 1998 (HSE, 1998)[7].

The HSE report said that the primary cause of the accident was the driver passing a red signal after going through two cautionary signals. However, the HSE also criticised Railtrack for failing to reposition a speed restriction sign on the main line between Euston and Scotland which could have averted the collision with an empty train. The HSE said that Railtrack failed to set up proper investigation procedures after four previous incidents of trains passing signals at danger at the accident spot. The inappropriate position of the speed restriction sign had given confusing information. The HSE report said that repositioning the sign, together with better brakes and the provision of an automatic train protection (ATP) system, which prevents drivers going through signals at danger, could all have prevented the accident.

One of the most interesting debates has been about the cost and benefits of installing ATP, a system that stops trains automatically if the driver misses a signal. From an economic perspective, the costs of installing the ATP system throughout the network appear to outweigh the benefits. The key question is whether the regulatory institutions and politicians have the courage to spell out the logic of a decision to proceed with a less expensive option, even accepting the argument that people expect railways (and airlines) to be safer than roads.

Roads

The sunk costs of the road industry are much lower than those experienced by the rail industry. Except where road tolling is in force, the users of roads are not directly faced with the full costs of their modal choice, and entry to the market has traditionally been easy. The available evidence suggests there are substantial weaknesses in the safety and quality regulation of road transport, which is in the hands of the police, the Traffic Commissioners and the Vehicle Inspectorate. During the period of regulated bus operations (1930-1986), quasi-judicial proceedings were held by the Traffic Commissioners to deal with applications to enter the market and to raise fares. These procedures were swept away with the Transport Acts of 1980 and 1985, leaving operators outside London free to operate routes and fix fares, subject only to general competition law.

Safety regulation in road transport involves approval of the type of vehicles, their maintenance and use and the road upon which they are driven. In the case of vehicle safety regulation, a system of operator licensing and maintenance arrangements is in place, backed up by annual inspections of vehicles and spot checks on the road. The Traffic Commissioners issue Operator Licences, and the entry requirements to the industry are not particularly stringent. In their inquiry into the adequacy and enforcement of regulations governing HGVs, buses and coaches, the Transport Select Committee reported one worrying problem to be the number of illegal operators in the road freight industry (HC, 1996)[8]. The witness from the Transport Development Group told the Committee that 'there could be 7,500 plus heavy goods vehicles running around our roads, unlicensed, untaxed or uninsured'. The Government recognises the substantial problem of unpaid Vehicle Excise Duty in the White Paper, and raises the issue of impounding unlicensed lorries (para 4.192).

The Transport Committee found that spot checks undertaken by the Vehicle Inspectorate and the police show low levels of compliance with regulations relating to overloading, hours of work, speeding and mechanical condition. Tachograph violation, in particular, was cited as the most common abuse of regulations in both the passenger and freight transport industries. With regard to overloading, convictions and prohibitions are growing at a rate of 5,000 to 6,000 per annum. This situation could be improved if lorries were weighed more regularly. On average, a vehicle is only weighed once every four years. In the case of mechanical condition, the DoT reported to the Committee that almost 5% of all accidents were caused by mechanical defects. The Vehicle Inspectorate estimate that a quarter of crashes they investigated were caused partly by brake failure. In October 1995 the Police found that more than 67% of lorries checked on the M2 in Kent had to be taken off the road for defects such as bald tyres, loose and missing wheel nuts and defective brakes.

With regard to the effectiveness with which regulations are enforced, the Committee reported that a constant complaint of witnesses was the leniency of penalties given by both the Traffic Commissioners and the courts against those who broke regulations. There was also a great deal of disquiet at the low level of fines imposed by magistrates on offenders, which were often not commensurate with the costs incurred by the Vehicle Inspectorate in bringing a prosecution or an effective deterrent. The former Senior Traffic Commissioner told the Committee that the maximum fine for illegal operation of an HGV was £2,500, but that the average fine imposed was only £300. The recommended fine in magistrates' guidelines for using a vehicle in a dangerous condition was only £600 even though the offence had a maximum fine of £5000. Better training of magistrates was suggested so that they would be aware of the seriousness of breaches of vehicle regulations. In addition, many witnesses drew attention to the fact that information about those found to be operating outside the law was not gathered together systematically so that the Traffic Commissioners could review the overall record of operators with a view to taking disciplinary action when appropriate. Institutionally, it is clearly necessary to raise entry standards and improve collation of information from spot checks to raise standards. There are plans to improve the co-ordination of information between Government and other agencies involved in enforcement activities (i.e. the Police, the Vehicle Inspectorate, the Driver and Vehicle Licensing Agency, the Traffic Area offices and local authority Trading Standards officers) using information technology through the Joint Enforcement Database Initiative.

Most industry participants want to see more resources devoted to enforcement, which they believe they pay for through their operator licence fees. The total cost of Vehicle Inspectorate enforcement work against

HGVs is around £11 million per annum and that of the O-licence system around £9 million. Since 1991-92, total staff numbers within the Vehicle Inspectorate have fallen by 17%. There has been a reduction in specialist traffic policing as more emphasis is given to local area policing. The Government has promised to bring forward a number of initiatives to promote higher safety and environmental standards in the freight transport industry. There are also plans to ensure that weighbridges are available for enforcement at all freight terminals and ports where this is justified by the levels of road freight traffic. Foreign-based lorries represent an enforcement problem because action against the licence holder takes place in the country of origin. Insofar as safe operation is concerned, highway maintenance standards are the responsibility of the Highways Agency and Local Authorities. There is strong evidence that Safety Schemes, even those with good rates of return, are not being carried out because of a shortage of funds, as is the case in Oxfordshire.

Competition

Railways

One of the underlying reasons for splitting railway operations and track ownership was the potential this provided for competition between railway operators over the same lines. However, protection was required to encourage companies to bid for the franchises at privatisation, and an incremental approach is now being taken to the introduction of competition because of concern that too rapid an introduction would erode network benefits. For example, without appropriate controls, unrestricted competition could lead to non-impartial service information or to unbalanced timetables as was seen following the deregulation of the bus services industry. The Regulator has already had to step in to ensure impartial information is made available to passengers under the existing Phase I system in which competition is very restricted, suggesting that this matter will need to be treated very seriously once Phase II begins.

Following a consultation on the development of the competitive framework for passenger rail services (ORR, 1997)[9], the Regulator published his conclusions in April 1998 that the next stage of competition would commence with the winter 1999 timetable. Very limited liberalisation is envisaged, which would include the following types of routes:

- airport links;

- services to leisure and shopping areas;
- cross-regional services.

Although new routes will be open to entry, train operating companies will continue to be protected on a substantial proportion of their franchise routes. For example, new services will be restricted from stopping at intermediate stations that provide a threshold level of the franchisee's overall revenue.

The Rail Regulator is currently undertaking a study of the likely scale and nature of new services, and of the type of criteria which should be applied when deciding whether to approve new services. Among the criteria to be considered are the size of the benefits of passengers on new services, the implications for franchises, and the impact on the budgets of the Director of Passenger Rail Franchising and the Passenger Transport Executives, which fund non-commercially viable but socially beneficial services.

What we have seen so far on the railways is a form of product differentiation whereby commuters can use a suburban train instead of an InterCity one. An example of this is between Peterborough and London. To some extent, however, InterCity had already done this in BR days by restricting the use of its trains by holders of discounted tickets like Travelcards. Because of the inflexibility of the railway timetabling process, competition is likely to be limited. The Regulator has already ruled against timing just ahead of rivals, and is probably well aware of practices that aim to take a handful of money from the revenue-pool without adding much to customers' convenience.

The real question is whether, with rolling-stock still in short supply, there will be much real open-access or scheduled competition. It is also questionable to what extent increased competition of this kind is necessary, bearing in mind the competitive pressure which already exists in many railway markets. The issue of quality may be much more important to most users.

Roads

Competition in road haulage and express and private hire coaches is unrestricted. In the bus industry, consolidation has carried on unabated, leading to domination by three major groups: Stagecoach, FirstGroup and Arriva, which each account for more than 12% of the market. A second tier of groups, including National Express Group, MTL Trust Holdings and

Go-Ahead Group each account for between 3% and 8% of the market. The remainder of the industry is split between small groups (two or three companies each) and single units, owned either by local authorities, company managers or company employees. Of note, however, is the steady growth of the 'independent sector', comprising operators owning less than 50 vehicles each.

The area where the regulatory authorities have been most heavily criticised is in local bus services. The costs of travelling by bus have increased in real terms more than the costs of motoring and, ten years after deregulation, fares were up 19% on average in real terms (Scott, 1995)[10]. It needs to be possible for small high quality entrants to challenge incumbents who have left gaps in their product lines or are simply overcharging, but those incumbents must also have immediate and effective protection from low quality entrants who launch a predatory attack. To date, the OFT has singularly failed to distinguish between high and low quality competition.

There is a strong case for limiting low-quality competition especially when established operators are being expected to provide vehicles with lower emissions and disabled access. With regard to the latter point, the Government has put forward proposals for public service vehicle accessibility regulation. The Traffic Commissioners also believe that entry standards need to be raised. It has been suggested that among the prerequisites of entry should be:

- a proper operating base;
- qualified maintenance staff;
- more stringent technical standards which will also cover emissions and accessibility (this will drive out low-quality vehicles);
- tougher operator licensing standards, including an upgraded certificate of professional competence;
- the submission of a 'business plan' which shows that provision is being made for depreciation, the employment of certified maintenance staff and the 'garaging' of vehicles overnight at operating centres;
- the ability to exercise some discretion in disallowing registrations which are clearly aimed at duplicating existing journeys rather than developing the timetable.

Where quality standards are met, the Traffic Commissioner should have discretion over service registration to ensure that headways are maintained which offer a spread of service. Those intervals should then be protected, perhaps through Quality Partnerships. Clearly predatory actions, such as offering free rides and swamping routes with buses, should be subject to immediate prohibition with financial penalties that hurt even the largest operators. Such provision has subsequently been made in the Competition Act 1998.

High-quality competition in the bus industry can be identified in places such as Glasgow, where Stagecoach has introduced high-quality express journeys from the suburbs which have challenged the slower 'all stops' services provided by the incumbent, FirstGroup. Stagecoach did not seek to duplicate the inner-city network of the incumbent operator and has thus avoided creating the congestion that brought about so much criticism of the competition which followed deregulation in 1985. This type of competition is real product differentiation and may be shown to have generated a small modal switch from cars. Competitive entry is the only real check on a lazy or greedy incumbent and the possibility must be maintained. It is also needed as a means of maintaining competition in the market for tendered services. This market is currently showing signs of decline. The Select Committee on Environment, Transport and Regional Affairs has called on the DETR to undertake a thorough examination of London Transport Buses' tendering arrangements (LTT, 27th August 1998). The Committee says that since late 1996 there have been an average of only two bids per tender, compared with six per tender in 1995, and 92.8% of tendered London buses are now run by just six major bus groups.

The alternative of Quality Contracts floated in the White Paper (para 3.20) envisages a situation where services, and presumably fares, are determined by a local authority, and an exclusive franchise is let for the operators of bus services in a defined area. In these circumstances there will be no competition on the road. I have grave doubts about whether sequestering the right of an incumbent to operate the services which have often been purchased will prove to be legally possible without substantial compensation. I am also concerned that the service pattern may be rigid, unimaginative and open to a great deal of political influence.

Institutionally, I am hoping that the Traffic Commissioners' powers will be strengthened and that the OFT, with new competition legislation at its back, will be able to take swift action to stop predation in its tracks.

Conclusions

I hope I have said enough to provide some response to the question posed. I do not believe the regulatory institutions as at present constituted are adequate to ensure delivery of land-based transport that meets the aspirations of users.

Notes

1 HC 286-I (1998) *The Proposed Strategic Rail Authority and Railway Regulation.* Third Report of the Environment, Transport and Regional Affairs Select Committee, Session 1997-8. The Stationery Office, London.
2 Booz Allen & Hamilton (1998) *Railtrack's Performance in the Control period 1995-2001*, Office of the Rail Regulator, London.
3 Financial Times, 25th May 1999.
4 Department of the Environment, Transport and the Regions (DETR) (1998), *A New Deal for Trunk Roads*, July.
5 Standing Advisory Committee on Trunk Road Assessment (1994), *Trunk Roads and the Generation of Traffic.*
6 HC 54-I (1995) *The Consequences of Bus Deregulation*, House of Commons Session 1995-96, HMSO, London.
7 HSE (1998) *Railway Accident at Watford*, The Stationery Office, London.
8 HC (1996) *The Adequacy and Enforcement of Regulations Governing Heavy Goods Vehicles, Buses and Coaches.* Fifth Report of the Transport Committee, Session 1995-96. July.
9 Office of the Rail Regulator (1997) *'New Service Opportunities for Passengers'*, October. ORR, London.
10 Scott, D. (1995) *'Local Bus Service in Great Britain: the Transport Act 1985 ten years on...'* Parliamentary Message.

10 Integrated Transport Policy: Implications for Regulation and Competition The July 1998 Transport White Paper and the 1998 Competition Act

MIKE HOBAN[1]

Introduction

This paper sets out to discuss the Transport White Paper[2] and the Competition Act 1998[3], with particular reference to local bus services. It begins with two arguable, preliminary positions. These positions are:- that the Director General of Fair Trading (DGFT) can, *inter alia*, be regarded as discharging a regulatory role; and that, with the Competition Act having received Royal Assent, the case is conclusively made that our current UK competition legislation is inadequate, and is now being addressed by that statute.

I turn first to the Director General as Regulator *Manqué*. The DGFT is currently empowered to intervene in markets directly. A Competition Commission (CC) reference[4] can result in the DGFT negotiating undertakings with the subject company to constrict the latter's activities in the relevant market[5]. Or those undertakings could be offered by a company *in lieu* of a reference to the CC[6]. In the transport industry such undertakings can amount to regulation targeted at transgressors, by limitations on fare rises[7]; maintenance of service quantity if a competitor enters the market[8]; or requiring access by other parties to an essential facility[9]. Breaches of such undertakings can be redressed by a Court Order[10] which, together with the attendant adverse publicity, act as a discipline upon transgressor companies.

It is also arguable, although with less evidence and therefore less conviction, that the Office of Fair Trading (OFT) exerts indirect regulatory effects on markets. For example, the actions of the OFT, and the conclusions of the CC in high profile bus predation cases, may deter other companies from pursuing such courses of conduct. The number of complaints of anti-competitive activity in the bus industry has been falling, and may be cited as evidence both of the deterrent effect of regulatory activity, plus other factors - notably, a maturing market. It should also be noted that there has been a diminution in merger activity in the bus industry (see Annex 1).

The very presence of the new legislation is the starkest evidence, however, that much of the structure and approach of the current competition legislative framework has outlived its usefulness. The exception is the Fair Trading Act 1973 (FTA), which remains on the statute book, as amended by the Competition Act 1998. The previous Government circulated, in August 1996, a draft Competition Act[11] - evidence of its apparent intent to change the law. The three main political parties, in their 1997 General Election Manifestos, pledged to introduce competition legislation[12]. The DGFT, in his 1997 Annual Report, commented that while the

> Fair Trading Act has proved to be an exceptionally useful and flexible piece of legislation, the Restrictive Trades Practices Act and the Competition Act are no longer effective in dealing with anti-competitive activity[13].

And that 1997 Annual Report was the tenth OFT Annual Report to refer to the need for competition reform[14]. The drawbacks of the current competition legislation are widely known - in the case of restrictive agreements

> too much time is spent in registering and scrutinising agreements of little or no competitive concern. Worse, investigatory powers to tackle cartels are inadequate[15].

In respect of anti-competitive practices in, for example, the bus industry, the House of Commons Transport Committee commented that 'we have received complaints that the OFT is slow both to initiate and conclude complaints and.. no compensation is awarded to a firm driven out of business as a result of unacceptable actions'[16]. The MMC report on the supply of bus services in the North East of England drew attention to the lack of interim relief for victims of predatory activity, and the cumbersome processes of current competition legislation[17]. Concerns over responses to alleged anti-competitive practices were also voiced in a Debate during the passage of the Competition Act in the House of Commons in May 1998[18]. And research has indicated that regulatory activity in bus markets has not generally led to third party entry and improved competition in the subject markets. Factors such as reputational effect are suggested as having played a role in inhibiting entry and competition[19].

Hence the weight of experience, and the views of past and current Governments, enables me to respond to the question posed by the theme. The current regulatory structure that the OFT and the other UK competition authorities[20] operate under is inadequate, and is now being comprehensively overhauled to meet the challenges of the future. I now move on to touch on the Competition Act, the Transport White Paper, and how the key provisions of that statute act on the measures announced in the Paper to improve bus services.

The Competition Act 1998

The OFT is engaged in a major education programme to draw industry's attention to the provisions of the Act, and the effect it will have on businesses throughout the UK. This includes guidelines, booklets, a video, radio tape, OFT speakers at events, business advice open days, an enquiry line and information on the OFT website (http://www.oft.gov.uk). The OFT has, so far, circulated nine Guidelines (with more planned), to meet the 'general advice and information' requirement of the Act[21] (see Annex 2). The key elements of the Act can be broken down into three parts, reflecting the first three Chapters of the Act.

The powers of investigation that the OFT possess are markedly strengthened by the Competition Act. Chapter I prohibits anti-competitive agreements. Chapter II bans the abuse of a dominant position. Chapter III sets out the investigation and enforcement powers to back up those prohibitions. The prohibition approach, modelled on Articles 81 and 82 (previously Articles 85 and 86) of the EC Treaty[22], is founded on the *effects* of an agreement, or of activity which abuses a dominant position. The Chapter I provisions prohibit agreements between parties which affect trade in the UK, and which may prevent, restrict or distort competition. It is the effect rather than the form of the agreement that is decisive. Chapter II of the Competition Act deals with the prohibition of abuse of a dominant position. It must be stressed that the prohibition is on the abuse of the dominant position, not on the actual holding of that position by a company. The Act sets out a series of illustrative activities that may constitute an abuse. Chapter III of the Act provides for the investigatory and enforcement powers for the DGFT under the new regime.

The July 1998 Transport White Paper

The July 1998 Transport White Paper was pre-figured in the Labour Party's 1997 General Election Manifesto. The new Government's objectives for transport were set out in more detail in the Department of the Environment, Transport and the Regions' (DETR) Consultation Document[23], produced in August 1997 in concert with the Scottish Office, the Northern Ireland Office and the Welsh Office. The Consultation Document generated a wealth of responses - over 7,300 written submissions, plus input from consultation meetings and seminars throughout the UK[24]. But the White Paper's central thesis - encouraging a shift from the private car to public transport, to reduce pollution and congestion - can be traced rather further back. Two White Papers in the mid-1960s - 'Transport Policy' published in July 1966[25]; and 'Public Transport and Traffic' published in December 1967[26], foreshadowed the Transport Act 1968. Both recognised the benefit of ready access to the car, but the 1966 White Paper counselled that:

> It has brought severe discomforts: congestion in the streets of our towns; ....noise, fumes and danger as the setting of our lives; a rising trend of casualties on our roads and a threat to our environment in both town and countryside[27].

The 1967 White Paper went on to comment that 'all the transport matters for which local authorities are to be responsible.... must be focused in an integrated transport plan'[28] - an echo, perhaps, of the local transport plans envisaged by today's White Paper[29]. And in the 1960s it was clear to Government that prioritising, deciding, and delivering public transport inputs 'must rest in the first place on local communities'[30]. Effective road space usage was important – 'to get more people moving with less use of road space is vital to the solution of our transport problems[31]. Back in 1967, 'full integration of the planning of road and rail services' was 'a pressing need'[32]. No real change in 1998 - the White Paper promises 'more through-ticketing; better facilities at stations .... for interchange; better connections between, and co-ordination of, services'[33].

The shift from private car to public transport found resonances closer to today, in the previous Government's Transport Green Paper of April 1996[34], which summarised one of the key themes of the 1995-6 Transport

Debate as a need to switch emphasis in spending from roads to public transport, with Central Government offering help to local authorities to deliver, *inter alia*, bus priority measures and local transport strategies[35]. And, in September 1997, a Royal Commission on Environmental Pollution (RCEP) report[36] remarked that the development of an integrated transport system required intervention by Government in the bus and rail industries. The RCEP also commented that there was a need for high quality public transport; and consistency between policies for public and private transport[37]. By the time the RCEP reported, Central Government had given a plain steer, through the Consultation Document of August 1997, about the course it intended to embark upon.

The 1998 White Paper wants a shift away from the private car to improved, integrated public transport. A mixture of 'carrots' and 'sticks' are proposed to deliver that. Carrots include a new fund to address capacity constraints on key parts of the rail network[38]; a national public information system accessible by telephone, teletext and Internet[39]; and local powers to ensure bus operators' participation in multi-operator ticketing schemes[40]. Sticks include powers for congestion charging[41], and workplace parking charges[42]. Revenue from those economic instruments will be channelled into transport improvements. The White Paper received a favourable reception - for example, of the 17 'key people' approached for comment by Local Transport Today, only one reacted negatively[43]. Seven Daughter documents, fleshing out policy in a variety of areas - including buses – have been published so far (see Annex 4). The Buses Daughter Document (BDD)[44] (released in March 1999) proposes, *inter alia*, that Quality Partnerships should be put on a statutory footing; that Quality Contracts could be introduced in particular geographic areas, subject to various tests; and that more effective timetable provision, passenger information, and through-ticketing be developed.

The literal interface between the Competition Act and the White Paper surfaces in five references[45]. But the most significant policy interplay between the White Paper and the Competition Act is in respect of local bus services. The OFT's response to the DETR's August 1997 Consultation Document concentrated on local bus services[46]. The rest of this paper follows that approach, limiting itself to bus services outside London and Northern Ireland, for a variety of reasons. Bus deregulation and the bus industry has rarely been far away from the public eye: witness the interest generated by the reference of Arriva's acquisition of Lutonian Buses to the MMC[47]. It is the transport mode that generates the most complaints within the Office (Annex 1). It is also one of the principal public transport modes in terms of passenger journeys, and vehicle kilometres. In 1996/97 there were 4,355 million passenger journeys on local buses in Great Britain[48] (compared to some 900 million train journeys, including London Underground trips[49]); and there were 2,693 million vehicle kilometres operated by local bus services in GB in 1996/97[50].

In addition, the growth in the number of Quality Partnerships (QPs) and the announcement on Quality Contracts raise competition issues. In respect of other transport modes, we would generally look to the Office of the Rail Regulator to take the lead on rail competition issues. The Competition Act 1998 gives the Rail Regulator concurrent powers with the DGFT in relation to the supply of rail services. Other transport modes do not generate significant numbers of complaints to the OFT. So, almost by a process of elimination, the OFT turns its primary attention to bus services in respect of the July 1998 Transport White Paper, and its relationship with the Competition Act.

Quality Partnerships and the Competition Act

The encouragement of QPs is one of the key elements of the White Paper. The TAS Partnership offered a definition of Quality Partnerships in 1995[51]. The RCEP also offered an explanation[52]. The White Paper offers a slightly longer, more descriptive, definition[53]. The thrust of the arrangement is that an operator provides better buses and services, and, in return, the local authority or Passenger Transport Executive (PTE) introduces improved infrastructure for QP bus operators' services in the area. At the time this paper was presented, about two dozen QPs were in operation or under development. By May 1999, that had risen to 43 QPs (see Annex 3). The emphasis in this part of the paper is, however, not on current QPs, but those which are now under development and are to be signed after the Competition Act's enactment.

The advantages of QPs are often quantifiable. A TAS survey[54] reported that 42% of local authorities and 58% of operators involved in QPs felt that QPs worked well in providing a mechanism for enhancing bus service quality and usage. Passenger growth figures passed to TAS[55] in 1997 reported 5% to 20% increases in ridership, after the implementation of a QP. The White Paper commented that QPs have increased patronage by 10%-20%, and by up to 40% with bus segregation and substantial improvements in infrastructure[56]. QPs have been developed in urban areas – for example, Birmingham and Edinburgh; and in rural districts, such as

Pwllheli[57]. QPs also offer collateral environmental benefits (such as low emissions, less congestion and pollution due to modal shift); and social benefits (low floor buses offer easy access for the elderly, disabled, and parents with buggies).

The detailed and complicated nature of the transitional arrangements to deal with the move to the Chapter I prohibition are discussed in the OFT Guidelines on transitional arrangements[58]. The attention of parties to QPs signed before Royal Assent is also drawn to the OFT's guidance on the bus industry and the Restrictive Trade Practices Act 1976 (RTPA)[59].

Those QP agreements to be signed between Royal Assent and commencement of the new law will benefit from a one year transitional exclusion from the Chapter I prohibition *after* its commencement, provided that the agreements do not involve price fixing, and not have an appreciable effect on competition such that it is appropriate for the DGFT to act before the end of the transitional period. For agreements signed in this period between Royal Assent and commencement, parties can apply for Early Guidance on Chapter I of the Act. That early guidance would expect to set out:- whether or not an agreement would be likely to infringe the Chapter I prohibition; and whether the QP agreement would be likely to receive an exemption, in the case of an infringement.

The Competition Act's main provisions are to come into force in March 2000. In the final analysis, the key issue under the Competition Act is whether or not a proposed QP might be in breach of the Chapter I prohibition on anti-competitive agreements. The short answer to the question 'Is my QP proposal compliant with the Competition Act's provisions?' is that each QP proposal would have to be considered on its individual merits against the prohibition and the criteria for exemptions set out in the Act[60] to establish whether that was the case. If a QP proposal is caught by the Chapter I prohibition, it could be considered for a possible exemption under the Act.

Before entering into general commentary on QPs and Chapter I, three issues should be borne in mind. First, legislation to give statutory footing to QPs is promised in the White Paper[61]. The BDD provides more detail on the form and nature of that legislation. However, the potential interaction between the Competition Act and the Transport White Paper's proposed legislation in this area is not yet known. Secondly, it may be the case that a local authority is not regarded as an undertaking[62] for the purpose of the Competition Act. However, each proposed QP would be scrutinised on its merits, and on the nature of the involvement of the local authority in that proposed agreement[63]. The final issue is that, given the variety of QP arrangements, in terms of activities undertaken and geographic areas, QPs are so diverse that it is questionable whether a block exemption under the terms of the Competition Act would be feasible[64], although the OFT will need to study a number of QP proposals before reaching a firm position.

In respect of proposed QPs, it should be borne in mind that, under the Competition Act and in line with European jurisprudence, only agreements that appreciably restrict, prevent or distort competition are prohibited. An agreement will be treated as appreciable if it has a significant adverse effect on competition, and the DGFT has stated in his guidelines that it will usually be assumed that there is no appreciable effect where the combined market share of the parties to the agreement are less than 25%, and the agreement does not involve price fixing, or market sharing[65]. In the case of proposed QPs therefore, the parties would only need to consider notification if their combined share of the relevant product and geographic market exceeded that level or if the terms of the QP otherwise appreciably prevented, restricted or distorted competition. It is only agreements that appreciably restrict competition that can benefit from exemption - other agreements fall outside the Act's scope. The key tests would be whether or not there was price fixing in a proposed QP; or an attempt to limit or control the market. The first step would be for the parties to the proposed QP to consult their advisers, and the relevant OFT guidelines, before reaching a decision on whether or not to notify the OFT.

During the period between Royal Assent (9 November 1998) and the Act's main provisions coming into force (1 March 2000) the OFT is able to offer Early Guidance to parties on whether or not their agreement is likely to infringe the Chapter I prohibition, once it comes into force; and, if so, whether it is likely to benefit from an exemption[66]. After the Act comes into force, in March 2000, notification may be made for guidance or a decision. Guidance may indicate whether or not the agreement would be likely to infringe the prohibition, and whether or not it would be likely to be granted an exemption if an application is made for a decision. A decision may be that the agreement is:- outside the prohibition; or is prohibited; or is exempt.

If it is decided to approach the OFT for confirmation that the agreement appreciably restricts competition and whether an individual exemption under the terms of the Competition Act was likely, the issue would then be progressed as follows. Four conditions - two positive, two negative - have to be met[67]. In respect of the positive conditions, the agreement has to have the effect of either:- improving productivity or distribution; or promoting technical or economic progress. It is arguable that QPs do improve productivity, and provision of services - buses may be able to move more quickly and efficiently within bus lanes, and progress more rapidly at traffic lights, as a consequence of upgraded infrastructure provided by local authorities.

To merit exemption the agreement must pass on a fair share of the resulting benefits of the agreement to consumers. In the case of QPs, modal shift effected by those arrangements may result in less congestion and easier movement of goods in urban areas, cutting delivery times and business costs. In terms of benefits to the customer, infrastructure improvements increase accessibility; improve punctuality; and increase the speed of journeys.

Two negative conditions also have to be satisfied, for the DGFT to be in a position to grant an exemption under the terms of the Competition Act. The first negative condition is that a QP should not impose on the parties' restrictions which are not indispensable to the attainments of the two positive conditions. Such restrictions would be curbs on either party that do not improve efficiency or promote economic progress. (The exemption procedures in the Act reflect the fact that some arrangements which may fall within the scope of the prohibition may in fact have wider economic benefits which more than offset their anti-competitive effect[68].)

The second negative condition that has to be met is to the effect that a QP agreement should not offer a chance for the parties concerned to eliminate competition. This would be the area of most concern to the OFT in respect of QPs. Standards set for QP entry should not be so high as to debar new entry by potential competitors to the QP. Thus far, given the existence of over forty QPs, the OFT has received no complaints from excluded bus operators unable to meet QP entry standards. The second concern is that a lower quality operator should not be prevented from running services in the geographic area of a QP if he takes the view that he does not want to make the requisite investment in, for example, low floor buses. If caught by the prohibition, a clause in a QP which sought to exclude a non-signatory from a particular locality or route would, depending on the relevant market definition, be in breach of the second negative condition, and an exemption could not be granted.

Exemptions will be granted subject to time limits and may be subject to conditions[69]. A subsequent material variation in a QP (for example, extension of a QP's geographic area) would need to be considered by the parties to the proposed agreement in the context of whether or not the question of appreciability arose, before they decided to notify the DGFT for the purposes of negative clearance or exemption.

The relationships between a QP and the RTPA, and the Chapter I prohibition are complicated, and sponsors of current and proposed QPs have to assess carefully the impact of that legislation on a QP agreement. Early Guidance on QPs and Chapter I may be sought from the OFT, if the parties consider it is warranted. In addition, the OFT will feed its comments on the relationship between QPs and the Competition Act into the OFT response to the DETR's BDD[70].

The White Paper supports QPs to the extent that new legislation to underpin QPs is promised. Evidence cited earlier indicates that QPs increase passenger numbers, and reduce congestion. The concern is that the ambition to facilitate modal shift vitiates the prospect of on-road competition to the extent that entry barriers are set too high; or low-cost potential entrants are banned, or restrictions are placed on the services they provide, leading ultimately to higher prices to customers. The aim must be to ensure that the potential friction between modal shift and competition is translated instead in to a complementary system whereby quality improvements are delivered, and passenger choice is enhanced. That can only properly be achieved by operators and local authorities comprehending the opportunities presented by the White Paper, and the values of on-road competition.

Quality Contracts

The White Paper indicates that a further power will be introduced through primary legislation to enable local authorities to enter into Quality Contracts (QCs)[71]. The QC concept is fleshed out in some more detail in the BDD[72]. A QC would be sanctioned by Ministers (or a devolved assembly) if a QP failed to deliver the necessary improvements in transport. In essence, a QC would be a franchise for an operator to run services on one route or a group of routes, without competition, for a certain period of time. This section of this paper discusses the benefits and disadvantages of QCs from a competition perspective.

The White Paper 'Buses' - which set out the case for bus deregulation - understandably did not regard franchising as the most effective way to exert competition on operators[73]. But bus franchising has been in operation in London since July 1985 when the first tenders within London were let by London Transport[74]. The advantages of QCs have been particularly espoused by the PTEs[75]. Benefits of QCs may be that the quality improvements sought by local authorities, but unachievable under a QP, would be made a key contractual aim of a QC. A stable network, compulsory involvement in multi-modal travelcard schemes, along with contracted investment, are regarded as other advantages of that arrangement.

But the QC concept has drawbacks. Competition in the market is preferable to competition for the market. There may be bureaucratic hurdles to overcome - a complaint made by bus company managers in London concerning tendering[76]. The extra costs which a QC imposes on a company in respect of investment may be passed on to customers through fare increases. Innovation may be thwarted by the attempt to procure stability, causing entrepreneurial frustration, route ossification, and a poor service to customers. Without the discipline of competition, there may be a risk that service quality may suffer, with a monopolist entrenched and possibly well equipped to see off potential entrants in a new tendering round.

The White Paper and the BDD implicitly appear to acknowledge that a decision to move to a QC is a difficult one to take. Not only would a QC require Ministerial (or a devolved assembly's) consent, but, in addition, pilot projects would first be established to probe the benefits and disadvantages of QCs. Guidance and a formal appraisal process for QCs would also be published[77]. 'Best practice', and Ministerial scrutiny should provide objective benchmarks that would help ensure that a QC is only sanctioned when it is clear that a QP is insufficient to guarantee necessary improvements, and it is in the public interest for a local authority to introduce a QC.

The OFT expects to be consulted by DETR as the process moves forward. And, in respect of fairness and transparency, a local authority, in considering a QC submission, would have to observe the 'Best Value' criteria in awarding a contract. Allegations of unfairness in awarding QC tenders could be investigated through the Local Government Ombudsman or the Audit Commission. The QC approach is a bold initiative - but the safeguards proposed by the White Paper and the BDD should be able to ensure that it is applied only where a conclusive case is made for its introduction.

Chapter II of the Competition Act

This paper now focuses on predation and the Chapter II prohibition[78]. In establishing whether the Chapter II prohibition would apply, two tests have to be met - firstly, whether the subject company is dominant; and, secondly, whether it is abusing that position. In assessing dominance, the DGFT has to define the market in which the undertaking is alleged to be dominant. For the market share of the alleged dominant player can only be properly determined after the boundaries of the market have been defined. The relevant market has two dimensions - the product market - the market for local bus services; and the geographic market - the actual geographic extent of the market. In the latter case, that might be the local authority area, or even a handful of routes. After market definition, the next step is to assess dominance. Essentially, a dominant market position is one which enables a company to behave independently of competitive pressures. The European Court of Justice (ECJ) defined such a position as: -

> ... a position of economic strength enjoyed by an undertaking which enables it to prevent effective competition being maintained on the relevant market by affording it the power to behave to an appreciable extent independently of its competitors, customers and ultimately of consumers[79].

This interpretation will apply under the Act, by virtue of section 60 of the Act, which essentially imports 35 years of EC jurisprudence[80]. In attempting to determine whether there is dominance, the Director General has to assess whether, and to what extent, an undertaking will face constraints on its ability to behave independently. These constraints could be described as -

- Existing competitors, and their market share;
- Potential competition; - the likelihood (or otherwise) signalled by lack of high entry barriers, and existence of potential new entrants; and
- Other constraints, such as strong buyer power.

In terms of price predation in the bus industry, the first two constraints - other bus companies, and low entry barriers - would usually be the sole determinants of a predator's ability to act independently. The DGFT has stated in his guidelines that an undertaking is unlikely to be dominant if its market share is below 40%[81].

Once market dominance has been established, the second key issue to be settled is whether there has been an abuse of the dominant position in the relevant market. That anti-competitive conduct can be characterised as a course of action that exploits customers or suppliers, or is intended to remove, limit or prevent competition.

Price predation is an example of anti-competitive conduct, and an activity that has spawned more than its fair share of OFT investigations in the past. Although consumers may benefit in the short term from lower prices, in the longer term higher prices, reduced quality, and less choice will result from the weakened competition in the market. It is often difficult to distinguish, however, predatory behaviour from aggressive competition. A distinction has to be drawn between predatory behaviour on the one hand, and lower prices which have resulted from legitimate competitive behaviour on the other. In the past, we have used the test in the OFT Research Paper on price predation[82]. With the move to the Article 81 and 82-based approach, the OFT will need to turn to EC precedents to drive its determination of abusive conduct *vis a vis* price predation. The two key European cases are *AKZO*[83] and *Tetra Pak II*[84]. In the *Tetra Pak* case, the ECJ followed its earlier decision in *AKZO*, where it confirmed that if prices are below the average variable cost of production, predation should be presumed. The ECJ also held that if prices are above average variable cost but below total average costs, conduct is to be regarded as predatory where it can be established that the purpose of the conduct was to eliminate a competitor.

Chapter III of the Competition Act

The final area of the Act this paper discusses is the investigatory and enforcement powers in Chapter III, which apply to all OFT enquiries into alleged infractions of the Chapter I and II prohibitions. One of the principal differences between the current regime of investigating scale and complex monopolies, anti-competitive practices and registrable agreements, and the future approach set out in the Competition Act, is that in the former, serious anti-competitive behaviour can continue unchecked throughout an investigation, sometimes for a lengthy period[85]. Competitors who have been damaged by the activity may be driven out of business before matters are found, too late, in their favour. This point was mentioned in the Commons Debate in May 1998 during the passage of the Competition Act[86]. Some of the key changes to the investigatory and enforcement regime act to confront those deficiencies. Firstly, the DGFT has the power to impose 'interim measures' during an investigation, compelling the suspension by the company under investigation of the alleged anti-competitive practices[87]. In other words, the DGFT can put a temporary stop on irrational low prices in a market, or other anti-competitive conduct. Those measures can be applied (through a written notice) when the DGFT has a reasonable suspicion that a Chapter I or Chapter II prohibition has been infringed. However, such measures can only be effected for the purposes of preventing serious, irreparable damage to a particular person; or to protect the public interest. The OFT's analysis would involve a careful study of the rights of both the complainant and the complainee.

The DGFT can, if he has 'reasonable grounds for suspecting' infringements of the Chapter I or II prohibitions[88], require the production of any specified document or specified information. He will have power to enter premises, unannounced where necessary. In addition, on the authority of a High Court (Court of Session in Scotland) warrant, he will be able to search premises and will be able to use such force as is reasonably necessary for entry, if the judge is satisfied that there are reasonable grounds for suspecting that the premises contain key documents that should be provided, but have not been provided on earlier occasions, or would be destroyed if requested, or the investigating officer has been unable to enter[89].

Under the terms of the current regime, monopolies and anti-competitive practices may be referred to the Competition Commission, and may be the subject of legally binding undertakings to curtail those activities. Or undertakings may be accepted *in lieu* of a reference to the CC. Under the new legislation, the DGFT, if he is satisfied that there has been a breach of the Chapter I or Chapter II prohibitions, may give a direction to the parties concerned to modify or terminate the agreement, and to refrain from entering into similar agreements; or to modify or cease the anti-competitive conduct[90]. Failure to comply with such a direction leaves the relevant parties exposed to damages by those affected by the decision, and the DGFT may seek a Court Order to enforce his direction[91]. (A claim for damages in respect of the *original* infringement could also be lodged by an affected party.) An agreement which infringes the Chapter I prohibition is void and cannot be enforced. Currently, there is no financial penalty attached to a finding of anti-competitive practice. That current inability to levy financial penalties contrasts with the fines imposed by the European Commission. For example, in the *Irish Sugar* case the subject company was fined 8.8 million Ecus[92]. Under the new legislation, however, a civil fine of up to a maximum of 10% of turnover may be levied on a company judged by the DGFT to have been in breach of a prohibition[93]. The DGFT's determinations are appealable to the CC[94].

The granting of these new powers to investigate and punish anti-competitive agreements and the abuse of dominant positions is not without precedent. The United Kingdom is currently the only European Union Member State with no powers to make on-site inspections. The new powers are significantly weaker than those in the United States and Canada, whose competition authorities possess criminal investigatory powers, such as the impanelling of a Grand Jury, and the use of FBI covert investigations, to discover criminal anti-trust violations. The OFT's new investigatory powers, and the ability to obtain interim relief, put a premium on time, and replace the current cumbersome and over-lengthy procedures. That can only be of benefit to operators and to local authorities keen to grasp the opportunities presented by the Transport White Paper and the BDD.

Conclusions

Since deregulation we have witnessed continuing concentration in the bus market, and less complaints about anti-competitive practices by bus companies passed on to the OFT (see Annex 1). The drop in ridership outside London is well documented[95], although the evidence cited earlier indicates[96] that current QPs are attracting passengers back on to buses. To revive the bus market, and to deliver the modal shift from private car to public transport that is the underlying aim of the White Paper, the Deputy Prime Minister called for the bus to be not only the workhorse, but the 'racehorse' of the transport world[97]. This is what the White Paper and the BDD seek to put into effect.

The development of QPs, and the interaction with the Chapter I prohibition, will be of key interest to the OFT, as it moves from Enactment to Commencement. The message is that if parties have any concerns about a proposed QP and the Competition Act, the OFT is prepared to offer informal advice on current legislation, and on the Competition Act.

It is Central Government's aim that the development of QPs, and the stronger powers placed in the DGFT's hands with the advent of the Competition Act, will combine to ensure that the benefits of competition flourish alongside QPs, and predatory action would be apparent not only to victims, but to local authorities, who would be expected to alert the OFT. The power of interim relief - to prohibit further alleged predatory or other anti-competitive activity until the completion of an investigation - would act as a discipline on those tempted to cross the frontier between keen competition and predation. The increased powers of investigation, and the large financial penalties available if an allegation is found to be proven, are signals that the OFT is unwilling to countenance anti-competitive behaviour in this changing market, as elsewhere.

Annex 1

Table 1 Complaints received on alleged anti-competitive practices on land-based public transport

	1992	1993	1994	1995	1996	1997
Bus operators	90	108	56	60	45	34
Railways	7	6	5	3	1	3
Taxis	15	15	10	4	5	14

Table 2 Mergers and proposed mergers considered (including management buy-outs and coach mergers)

Year	Total No. of Mergers considered	No. of cases found not to qualify	Qualifying Mergers		
			Cleared	Referred MMC	Undertakings in Lieu
1990	18	6	7	5	0
1991	8	5	3	0	0
1992	2	1	1	0	0
1993	13	6	5	2	0
1994	24	9	12	3	0
1995	35	6	25	4	0
1996	31	11	17	2	1
1997	13	5	8	0	0
1998	14	3	10	1	0
Total	**158**	**52**	**88**	**17**	**1**

Source: OFT Statistics NB Figures do not include cases abandoned.

Annex 2

OFT Guidelines as of June 1999

Title	Current Position
The Major Provisions	Published March 1999
The Chapter I Prohibition	Published March 1999
The Chapter II Prohibition	Published March 1999
Market Definition	Published March 1999
Powers of Investigation	Published March 1999
Concurrent Application to Regulated Industries	Published March 1999
Transitional Arrangements	Published March 1999
Trade Associations, Professions and Self-Regulating Bodies	Published March 1999
Enforcement	Published March 1999

Public Consultation Completed

Mergers and Ancillary Restraints
Assessment of Market Power
Assessment of Individual Agreements and Conduct
Application in the Telecommunications Sector

Consultation Drafts in Preparation

General Economic Interest
Application in the Energy Sectors
Vertical Agreements and Restraints
Intellectual Property Rights
Land Agreements
Application in the Water Sector

Annex 3

Bus Quality Partnerships

Madingley Road P&R, Cambridge
Newmarket Road P&R, Cambridge
Central Leicestershire

Loughborough
Gwynedd
Easy Access 480 (Gravesend/Dartford)
Stafford Highfields
Northampton Northern Corridor
Perth - Tulloch
Worthing
Coventry Showcase
Coventry Showcase – Foleshill Road
Salford/Hope Hospital
Next Bus Clipstone
Super Route 88 (Ipswich)
Chantry - Whitehouse (Ipswich)
Super Route 66 (Ipswich)
Scott Hall Road (Leeds)
West Derby Road (Liverpool)
Brighton Network
Pheasey Corridor Showcase (Birmingham)
Cleveleys - Martin Mere (Blackpool)
Stoneydelph (Tamworth)
Sunderland
Culter Corridor (Aberdeen)
Norfolk County Partnership
Hull City
Greater Manchester PTE (proposed)
Hinckley
Pwllheli
Flintshire
North Tyneside
Newcastle City
Swansea
Bristol
Preston
Edinburgh "Greenways"
Norwich Western Corridor
Oxford
Nottingham: Derby Road; Mansfield Road; Hucknall Road
Colchester Service 65

Source: Confederation of Passenger Transport   May 1999

Annex 4

Daughter Documents From The Transport White Paper 'A New Deal For Transport: Better For Everyone'

Railways Policy: a response to the third report of the Environment, Transport and Regional Affairs Committee on the proposed Strategic Rail Authority and Railway Regulation (7/98)

Trunk Roads policy: Outcomes of the reviews for England and Wales (8/98)

Charging Policy - a consultation paper on implementing road user charging and workplace parking charges (12/98)

Shipping Policy – a response to the recommendations of the Working Group on Shipping – Charting a new course (12/98)

From Workhorse to Thoroughbred – a better role for bus travel (3/99)

Freight Policy – sustainable distribution: a strategy (3/99)

Guidance on Local Transport Plans (4/99).

Notes

1 The views expressed in this paper are personal, have not been adopted or approved by the DGFT, and should not be relied upon as a statement of the - views of the DGFT.
2 *A New Deal for Transport: Better for Everyone.* The Government's White Paper on the Future of Transport. Cm 3950, July 1998.
3 The Competition Act completed its Parliamentary stages later in 1998, and received Royal Assent on 9 November 1998. Its principal provisions are to come into force in March 2000.
4 The Competition Commission has succeeded the Monopolies and Mergers Commission - see section 45 of the Competition Act 1998.
5 Section 88, Fair Trading Act 1973.
6 Section 56A, Fair Trading Act; section 4, Competition Act 1980.

7   Monopolies & Mergers Commission (MMC) Report - *Southdown Motor Services Limited.* Cm 2248, June 1993. Undertakings announced in Department of Trade and Industry (DTI) Press Notice, 27 July 1994.

8   MMC Report - *The Supply of Bus Services in the North East of England.* Cm 2933, August 1995. Undertakings announced in DTI Press Notices on 8 October 1996 and 3 April 1997.

9   OFT Report - *The Southern Vectis Omnibus Company Limited.* February 1988. Undertakings given in lieu of reference to the MMC. Undertakings announced in OFT Press Release, 6 May 1988.

10   Section 93A of the Fair Trading Act.

11   *Tackling Cartels and the Abuse of Market Power: A Draft Act.* DTI, August 1996, which focused mainly on possible Article 81 (ex Article 85) Treaty type legislation.

12   *You Can Only Be Sure with the Conservatives* - the Conservative Party Manifesto 1997. *Make the Difference* - the Liberal Democratic Manifesto 1997. *New Labour - Because Britain Deserves Better* - Labour Party Manifesto 1997.

13   *OFT - Annual Report of the Director General of Fair Trading, 1997.* June 1998.

14   OFT Press Notice of 29 June 1998.

15   A *Prohibition Approach to Anti-Competitive Agreements and Abuse of Dominant Positions: Draft Act.* DTI, August 1997.

16   House of Commons Transport Committee Report *The Consequences of Bus Deregulation.* Paragraph 142, Volume 1. November 1995. It should be borne in mind, however, that civil actions are possible under section 35(2) of the Restrictive Trades Practice Act 1976: and section 93 and 93A of the Fair Trading Act 1973 (for both Competition Act 1980 and FTA undertakings and orders).

17   Paragraphs 2.158 and 2.164.(d) of MMC Report *The Supply of Bus Services in the North East of England.*

18   See columns 25, 27-8 of House of Commons Hansard, 11 May 1998, Volume 312, No.169.

19   *The Effectiveness of Undertakings in the Bus Industry.* OFT Research Paper No.14, prepared for the OFT by National Economic Research Associates, December 1997.

20   The OFT, the Secretary of State for Trade & Industry, the Monopolies & Mergers Commission (now the Competition Commission), and the Restrictive Practices Court are the relevant regulatory authorities under the current regime.

21   Section 52.

22   Treaty Establishing the European Community (Treaty of Rome), March 1957. See page 441, Butterworths Competition Law Handbook, 5th edition. The

Treaty of Amsterdam, which amended the Treaty on European Union, the Treaties establishing the European Communities (including the Treaty of Rome), and certain related acts, was signed at Amsterdam on 2 October 1997, and came into effect on 1 May 1999 (OJ C340, 10.11.97, p.1).

23 *Developing an Integrated Transport Policy: An Invitation to Comment.* August 1997.

24 See Annex B to the 1998 Transport White Paper.

25 Cmnd 3057.

26 Cmnd 3481.

27 Paragraph 1 of 1966 White Paper.

28 Paragraph 10 of 1967 White Paper.

29 Paragraphs 1.26-9, and 4.73 of 1998 White Paper.

30 Paragraph 11 of 1967 White Paper.

31 1966 White Paper.

32 Paragraph 17 of 1967 White Paper.

33 Paragraph 3.48 of 1998 White Paper.

34 *Transport - The Way Forward. The Government's Response to the Transport Debate.* Cm 3234, April 1996.

35 Page 8, Transport Green Paper.

36 The RCEP's 20th Report. *Transport and the Environment - Developments since 1994.* Cm 3752, September 1997.

37 Paragraph 8.29 of the RCEP Report.

38 Paragraph 4.32 of 1998 White Paper.

39 Paragraph 3.74 of 1998 White Paper.

40 Paragraph 3.55 of 1998 White Paper.

41 Paragraph 4.94 of 1998 White Paper.

42 Paragraph 4.107 of 1998 White Paper.

43 Local Transport Today, 30 July 1998.

44 Available on DETR's website (http://www.detr.gov.uk) or from DETR Free Literature, PO Box No. 236, Wetherby LS23 7NB.

45 Paragraphs 2.33-37 deal with maintaining competition, regulating monopolies, and bus deregulation. The interface also arises at paragraph 3.68 (service co-operation and the Competition Act); paragraph 4.23 (Rail Regulator to have concurrent powers with DGFT in relation to the supply of rail services under the Act); paragraph 4.26 (Act as a protecting mechanism against possible abuse of market position by ROSCOs); and paragraph 4.41(proposal for CAA to have concurrent powers with DGFT under the Act). There is also a reference to the OFT's Report on *Selling Second Hand Cars* - see paragraph 5.7.

46 *Developing an Integrated Transport Policy: OFT Response to DETR Paper.* November 1997, deposited with DETR on 12 November 1997. Copies available from author of this paper.

47 Independent and Daily Telegraph, 11 July 1998; Transit 27 July 1998; Coach and Bus Weekly, 16 July 1998. In the end, the MMC found against the merger – see *Arriva plc and Lutonian Buses Ltd: A Report on the merger situation*, 11/98, Cm 4074, ISBN 0-10-140742-4.

48 *Bus and Coach Statistics, Great Britain 1996/97*. HMSO, October 1997.

49 *Social Trends 28*. Table 12.14. HMSO, 1998.

50 *Bus and Coach Statistics, Great Britain 1996/97*.

51 *Quality Partnerships in the Bus Industry: A Survey and Review*. The TAS Partnership Limited, September 1997.

52 Paragraph 4.51, 20th RCEP Report.

53 Paragraph 3.16, 1998 White Paper.

54 Paragraph 5.2.5 of *Quality Partnerships in the Bus Industry*.

55 Paragraph 5.6.5 of *Quality Partnerships in the Bus Industry*.

56 See green box on page 40 of the 1998 White Paper.

57 See page 22 of the BDD.

58 *A Guide to the Provisions of the Competition Act 1998 - Transitional Arrangements* July 1998.

59 *Restrictive Trades Practice in the Bus Industry* OFT, March 1995.

60 Section 9.

61 Paragraph 3.17, 1998 White Paper.

62 Note the use of the term 'undertakings' - and not 'firms' or 'companies' in the Act. The term 'undertaking' has long been understood under European Community law to apply to any entity or natural person that carries on activities of an economic nature, including non-profit making activities. 'Undertaking' is often taken to have the widest possible definition - to catch as many persons as possible. In the report on Distribution of Package Tours During the World Cup (OJ [1992] L 326/31), for example, the EC Commission held that FIFA, the Italian FA, and the local organising committee were all 'undertakings'.

The handling of whether a local authority as a party to a QP is acting as an undertaking is also dependent on the legal basis on which QPs are eventually statutorily established.

63 The term 'agreement' has a similarly wide definition - prohibition can cover agreements whether legally enforceable or not, written or oral, and includes so-called 'gentlemen's agreements'. There does not have to be a physical meeting of the parties for an agreement to be reached: an exchange of letters or telephone calls may suffice if a consensus is arrived at as to the action that each party will, or will not, take. The prohibition covers anti-competitive behaviour between undertakings that can occur without a clearly delineated agreement.

It should also be borne in mind that whereas under the RTPA two or more firms have to accept restrictions before an agreement becomes registrable, the

Competition Act could prohibit an agreement between undertakings that contains restrictions on the activities of only one undertaking.

64 Annex 3 sets out the QPs that the Confederation of Passenger Transport understands are in operation. The numbers range from a handful of buses to some 1500 in the planned Greater Manchester PTE QP.

65 See paragraph 2.18 *et seq* of *The Competition Act 1998: the Chapter I Prohibition*, March 1999, OFT 401.

66 Further details can be found in *The Competition Act 1998: Transitional Arrangements*, March 1999, OFT 406.

67 Section 9.

68 Pursuant to section 60 of the Competition Act 1998, the exemption provision in section 9 of that Act can be interpreted broadly, and can involve the weighing of any anti-competitive effects of an agreement with environmental, social and other benefits.

69 Section 4.

70 See Annex A to the 1998 White Paper.

71 Paragraph 3.20 of the 1998 White Paper.

72 See Chapter 6 of the BDD.

73 Paragraph 37 of Annex 2 to *Buses*, Cmnd 9300, July 1984.

74 Page 15, *London Bus Tendering* Greater London Group, London School of Economics (D Kennedy, S Glaister and T Travers).

75 See the Passenger Transport Executive Group's publication, *Better Buses* (undated).

76 See page 151, *London Bus Tendering*.

77 See paragraph 6.8 of the BDD.

78 Guidance on Chapter II can be found in *The Competition Act 1998: the Chapter II Prohibition*, March 1999, OFT 402.

79 Case 27/76 *United Brands and United Brands Continental BV v EC Commission*, [1978] ECR 207, [1978] CMLR 429.

80 Subject to a number of provisos, the DGFT must act with a view to securing that there is no inconsistency between the principles applied and decisions reached under Part I of the Act (which includes the Chapter I and II prohibitions) and the principles laid down by the Treaty of Rome and the ECJ; and any relevant decision of that Court as applicable at that time in determining any corresponding question in EC law.

81 Paragraph 3.13, OFT Guideline, *The Chapter II Prohibition*, OFT 402, March 1999.

82 *Predatory Behaviour in UK Competition Policy*. OFT Research Paper No. 5. November 1994.

83 *AZKO Chemie BV v Commission* [1991] I ECR 3319; [1993], 5 CMLR 215.

84 *Tetra Pak II* [1997] 4 CMLR 662.

85 It should be noted, however, that section 3 of the RTPA currently allows for interim orders.
86 See above, endnote 18.
87 Section 35.
88 Section 25.
89 Section 28.
90 Section 32 and 33A.
91 Section 34.
92 OJ L 258, 22 September 1997 (page 1). EC XXVIIth Report on Competition Policy, 1997.
93 Section 36.
94 Section 46.
95 Table 2.1, *Bus and Coach Statistics* Great Britain, 1996/97.
96 See paragraph 15 above.
97 House of Commons Hansard, Debate of 20 July 1998, column 785.

11 Regulation of Bus and Rail: Is the Current Framework Adequate?

PHILIP O'DONNELL

Introduction

In their chapters, Mike Hoban and Bill Bradshaw provide different perspectives on the regulation of the privatised bus and railway industries. Mike Hoban deals with the regulation of bus services within the context of developments in UK competition policy. Bill Bradshaw's paper draws attention to significant differences in the approach which the Government has adopted to the regulation of the bus and rail industries and the weakness of regulation.

In my comments, I also seek to draw out the differences in the approach adopted to the regulation of the bus and rail industries, including on liberalisation of competition. I relate these differences to the privatisation models adopted for the two industries. I also outline how the policies set out in the Government's July White Paper, *A New Deal for Transport*[1], seek to address perceived weaknesses in regulation.

Regulation of Bus and Rail

The previous Government adopted very different approaches to the regulation of bus and rail. The bus industry was deregulated when the privatisation programme was still in its infancy. Public sector bus companies operated alongside a significant private sector. Rail privatisation was one of the last to be completed by that Government and the approach adopted reflected experience acquired over the previous years. The framework established for the regulation of local bus services was deliberately minimalist. The licensing regime continued to focus on safety. No industry specific regulator was established. The approach reflected:

- the perceived success of the earlier privatisation of long distance coach

179

travel which had led to lower fares, an expansion of services and greater use;

- the view that local bus markets would be contestable with no significant entry or exit barriers;
- the desire to transfer economic decision making for local bus services to the private sector; and
- the expectation that efficiency savings spurred by deregulation would allow outputs to be preserved while subsidy would be reduced.

Local government was to be left in the peripheral role of supporting non-commercial bus services, but with no controls over services which operators were able to provide without subsidy.

The regulatory framework adopted for the privatised rail industry is in marked contrast to that for local bus services. Existing British Railways scheduled passenger services, other than European Passenger Services, were franchised and subject to both contract and licence regimes to protect quality. The Railways Act 1993[2] established the Franchising Director with statutory responsibilities to secure the continued provision of franchised services. The Franchising Director has powers to regulate fares, to protect the level and quality of franchised services and to protect certain 'network benefits' such as interchangeable tickets between franchise operators, and multi-modal ticket schemes between trains and local bus services. The Act also established the Rail Regulator with powers to licence train operators and to require their co-operation on matters such as through-ticketing (another network benefit). The Regulator approves track access contracts between Railtrack (the network infrastructure provider) and train operators, including track access charges and the terms of on-rail competition.

The reasons for the different approaches adopted in bus and rail are not hard to identify:

- dissatisfaction with the perceived detrimental effects of the privatisation and deregulation of local bus services gave rise to concern that service provision would suffer as a consequence of rail privatisation, unless the Government took action to protect passenger services;
- a monopoly infrastructure provider (Railtrack) was being created and subsequently privatised. The customers of this new network utility, the passenger and freight train operators, required protection;
- Government, having accepted commitments to protect passenger services and facing a substantial on-going support payments to franchise operators, wanted a regulatory regime to protect the interests of passengers and the tax payer; and

- the perceived need to ease the franchising sales process and reduce the costs of franchising by moderating competition.

In the light of subsequent experience there appears to be widespread agreement that neither of the approaches adopted to regulation has proved satisfactory. Referring to local bus services, Bill Bradshaw notes that fares are reported to have risen by 19% on average in real terms in the ten years after deregulation. There have, however, also been offsetting reductions in the costs of public support. The White Paper, *A New Deal for Transport,* records that bus deregulation outside London has caused substantial upheaval because of bus wars and confusion over changing service patterns. While the White Paper recognises there have been some good examples of innovation, frequent changes to bus timetables, poor connections and the reluctance of some operators to participate in information schemes and through-ticketing arrangements have undermined bus services.

The regulation of the railways is also widely seen to be flawed. Bill Bradshaw refers to concerns about Railtrack's lack of commitment to undertake enhancement investment, describing Railtrack as extremely risk averse. He argues that the railways are locked into a system where Railtrack and franchisees decide what investment takes place, and the regulatory institutions are in a relatively weak bargaining position. He notes there has been concern about the quality of railway services. The annual report of the Central Rail Users' Consultative Committee (CRUCC) recorded a rise of 103% to 19,972 in the number of complaints received during the year to 31 March 1998 compared to 1996 - 97[3].

The White Paper also draws attention to the rise in complaints and to the comments of the CRUCC that 'there is a gulf between what passengers can reasonably expect and what they receive and how it is delivered'[4]. There is also reference to 'weaknesses arising from the fragmentation of the industry'[5] and 'there being no good ... mechanism for long term planning in the privatised industry'[6]. The House of Commons' Environment, Transport and Regional Affairs Select Committee is equally critical of the quality of services and also suggests that the overlapping responsibilities of the Rail Regulator and the Franchising Director has led to confusion about respective roles and inadequate sanctions on train operators and others who perform badly[7].

Regulation and Integrated Transport Policy

The White Paper sets out the Government's approach to regulation. A key consideration is that only by improving the quality of bus and rail services will it be possible to achieve Government's aim of attracting car users to make greater use of public bus and rail transport.

Significant improvements in bus services have been achieved through co-operation between local authorities and bus operators under Quality Partnerships. As Mike Hoban describes them, in Quality Partnerships local authorities provide traffic management schemes which assist bus services (bus lanes, priority at junctions, park and ride) while the bus operator offers better quality (in terms of comfort, greenness, accessibility and staff training), improved marketing, better integration and more reliable services.

The White Paper sets out Government's intention to tighten quality regulation. Government proposes bringing forward legislation to put Quality Partnerships on a statutory basis. This will give local authorities powers to require operators to meet certain standards of service quality in order to use the facilities provided by the local authority as part of the Quality Partnership. Local authorities will be better able to influence the provision of bus services and their marketing, and encourage the provision of easy access buses. The Government intends to clarify local authorities' powers to buy in extra services to boost frequencies on a particular corridor.

In some circumstances Quality Partnerships may not be enough to guarantee improvements and the Government therefore plans to introduce primary legislation to give powers to local authorities to enter into Quality Contracts for bus services. This would involve operators bidding for exclusive rights to run bus services on a route or group of routes, on the basis of a local authority service specification and performance targets. There may also be investment requirements. The Government's proposals are driven by the view that the market will not offer the quality of service and facilities necessary to attract them to public transport.

The Government also intend to take powers to enable the future Strategic Rail Authority (SRA) and the Rail Regulator to take action against operators who breach their franchise contracts, and to provide stronger and more accountable regulation. Again the concern is to achieve greater quality. In addition, the SRA is intended to provide a focus for the strategic planning of both passenger and freight rail services. It will ensure that the railways are planned and operated as a coherent network, not merely as a collection of different franchises, take a view of the capacity of the railway, assess investment needs, and identify priorities where operators' aspirations may conflict with one another.

The Role of Competition in Bus and Rail Industries

Mike Hoban and Bill Bradshaw offer contrasting views on the role of competition in the local bus industry. Bill Bradshaw argues that it is necessary to distinguish between high and low quality competition. Mike Hoban is more circumspect about how far it is appropriate to constrain low price low quality competition in order to encourage higher service quality. He recognises that there is an implicit tension between the general thrust of competition policy and the concern evidenced in *A New Deal for Transport*, to make public transport more attractive to the car user by driving up standards.

Mike Hoban records the views of past and present Governments that the existing competition powers which the Office of Fair Trading and other UK competition authorities exercise are inadequate. The Government's competition legislation, seeks to strengthen the hand of the competition authorities by prohibiting anti-competitive agreements, abuse of a dominant position, and creating investigation and enforcement powers to reinforce these prohibitions. The legislation is expected to come into force in March 2000.

Mike Hoban points out that experience with the deregulated local bus industry has served to highlight some of the weaknesses of the previous legislation. The industry has been subject to frequent OFT and MMC investigations following complaints about the competitive behaviour of bus companies including predatory pricing, and denial of access to key facilities. The MMC has also drawn attention to the lack of interim relief for victims of predatory behaviour, and the cumbersome processes of current competition legislation.

All this is a far cry from the original expectation. Competitive behaviour, instead of bringing the usual benefits, is thought by many observers to have brought no long-term gains to the bus passenger. Local authorities have continued to point to the loss of control and deterioration in quality and coverage of bus services, in particular the loss of quality timetables and lack of investment. Thus experience in the bus industry has highlighted some of the weaknesses in previous competition legislation and also led to the widespread perception that unrestricted competition is likely to compromise the ability to deliver raising quality standards and to the view that some restraint is necessary.

Mike Hoban recognises that Government's plans to develop Quality Partnerships and Quality Contracts raise competition policy issues. It is as yet unclear whether or not OFT would grant exemptions for Quality Partnerships under the new Competition Act - necessarily such exemptions would be case specific. He outlines the tests OFT would apply in determining whether an exemption is granted:

- does the partnership improve productivity or distribution; or
- promote technical or economic progress;
- does it avoid restrictions which are not indispensable to the attainment of the two positive conditions; and
- does the agreement provide the opportunity to the parties to eliminate competition.

He notes that the OFT would not expect standards for a quality partnership to be set so high as to debar new entry by potential competitors to the Quality Partnership. Nor should a low quality operator be prevented from running services if it takes the view that it does not wish to make the requisite investment in better quality vehicles. Underlying these points is a concern that facilitating modal shift through Quality Partnerships and Quality Contracts may frustrate on-road competition to the extent that entry barriers are set too high; or low cost potential entrants are banned, or restrictions are placed on the services they provide, and that this will lead ultimately to higher prices for consumers.

In discussing the benefits of the proposed Quality Contracts from the competition perspective Mike Hoban takes the view that competition in the market is preferable to competition for the market. His reasons are that lack of market competition may stifle innovation, lead to extra costs of administration and to higher costs including costs of investment being passed on to customers. Further, without the discipline of competition service quality may eventually suffer, if an incumbent franchisee becomes well entrenched and is able to see off competition in a new tendering round.

Bill Bradshaw's perspective on competition is different. While acknowledging that competition is the only real check on the monopolist, he argues that OFT has 'singularly failed to distinguish between high and low quality competition'[8]. He believes that the experience of the local bus industry demonstrates that there is a 'strong case for limiting low-quality competition especially when established operators are being expected to provide vehicles with lower emissions and disabled access'[9] and he draws a distinction with high quality competition offering real product differentiation and the prospect of modal shift.

The argument for moderating competition in bus services appears to rest on the proposition that market failure in the provision of road space, especially in congested areas, and in the consumption of scarce environmental resources, where road users do not pay their marginal social costs, justifies second-best policies to correct the market failure. These policies include emission standards, provision of dedicated facilities which can either be paid for directly by the taxpayer or through the fare box by limiting access. While the first-best economic solution would be to ensure that all resources used in road transport, including congested road space and environmental resources, are charged for at marginal social cost leaving operators free to compete including on price, there are a number of difficulties with the application of a first-best solution. The arguments for road pricing are considered in the White Paper. Pilot schemes are proposed.

The approach to competition adopted in the privatised railway industry stands in stark contrast to the approach so far adopted in the bus industry. Professor Bradshaw notes that while 'one of the underlying reasons for splitting railway operations and track ownership was the potential this provided for competition between railway operators over the same lines......protection was required to encourage companies to bid for franchises at privatisation'[10]. He also notes that there was concern that competition would erode network benefits. For example, without appropriate controls unrestricted competition would lead to unbalanced timetables as was seen following bus deregulation.

The Rail Regulator's December 1994 policy statement *Competition For Railway Passenger Services*[11] recognised the concerns to which Bill Bradshaw refers. In explaining the reasons for concluding that franchise operators should enjoy an initial period of exclusivity until 1999 he notes that:

- the franchising process itself creates considerable potential for increased competition even without allowing for "open access" operators. This reflects the geographic overlap between train operators in the way that the franchise map has been drawn; and

- in the early stages of privatisation it is difficult to predict the effects of allowing unrestricted - or uncontrolled - competition on train operators, the Franchising Director and passengers.

In June 1998 the Regulator published a consultation document *New Service Opportunities For Passengers - Criteria For Evaluation*[12] detailing his proposed approach to evaluating proposals for new rail services that will be able to operate when the current restraints are relaxed in 1999. He continues to take a cautious approach, being concerned to avoid the destructive elements of competition. The Regulator argues that 'It is essential that [liberalisation of access] takes place with appropriate safeguards. These need to take account of the special nature of the market, and to uphold the public interest in securing an integrated railway where services supported by taxpayers are not undermined and network benefits are protected'[13]. The Regulator looks to the development of competitive services focusing on the operation of new routes and new direct connections, subject to a public interest test taking into account the benefits to passengers, the impact on train operators and the impact on the Franchising Director's budget. The consultation document makes clear that the Regulator will not look favourably on proposals for new access rights which are aimed at abstracting business from existing operators. He will take into account the potential adverse impact of new services on the Franchising Director's budget and the Franchising Director's ability to continue to support broadly the current network. He proposes an assessment of the costs and benefits of new services as a basis for reaching conclusions. Competition will thus be subject to a merits test.

The different approaches to competition in the bus and rail industries reflect differing Government objectives for the two industries at privatisation. The measures the Government is now contemplating in the White Paper would bring the approach to competition in local bus services more into line with the approach to rail. Market protection will be considered as a means of improving service provision beyond the level which could be sustained in an entirely unregulated market.

The White Paper states 'that while competition can bring benefits to some customers, as suppliers compete for market share the wider public interest must always be taken into account. In transport the problems of noise, congestion and pollution associated with individual travel decisions may be ignored and there is a concentration on profitable routes at the expense of integrated transport networks which extend choice and accessibility'[14]. The White Paper offers a framework which:

- retains competition in the market but provides for intervention where there is evidence that this is needed in the public interest; and
- improves planning in a way which recognises the interactions between transport modes, land use and economic development.

In respect to rail, the White Paper sets out the Government's view that open access with inadequate safeguarding of the public interest could lead to a loss of network benefits like ticketing and timetabling. 'Cherry picking of the most profitable routes could threaten the Franchising Director's ability to support the current level of services. The Government expects the future Strategic Rail Authority to set the longer term policy for competition, ensuring safeguards against the erosion of a properly integrated network'[15].

The Nature of Privatisation: Rail and Bus Privatisation Contrasted

An important distinction can be drawn between the nature of privatisation in the bus and rail industries. While the arrangements for bus privatisation allowed for some continued financial support for services, the newly privatised businesses were left free to make their own economic decisions on output investment and fares subject to the constraints of a competitive market and passengers' willingness to pay. The approach taken to the privatisation of rail was different.

The vertical unbundling of British Rail had its origin in the previous Government's decision to continue to underwrite the provision of (passenger) rail services while seeking to contract out their provision by the most effective means. The solution adopted allowed for a high degree of market testing by breaking the industry down into many constituent components which were expected to contract with each other and with Government on a competitive basis. In effect, the Government looked to a mechanism to open up the industry to the pressures of a competitive environment while continuing to underwrite outputs. Privatisation therefore also represented an exercise in contracting out.

In commenting upon the subsequent performance of the rail industry, Bill Bradshaw questions whether the revised industry structure, with the separation of train operations from the provision of the track, and the creation of many train operating companies is fundamentally flawed and not capable of delivering the quality and efficiency of a well run vertically integrated railway. The perceived poor performance of the railway post privatisation may stem from several factors including:

- as Bill Bradshaw suggests, the form chosen for privatisation of the rail industry (vertical and horizontal unbundling into a number of companies);
- the robustness of the franchising contracts established between the Franchising Director and the franchisees;

- the robustness of the track access, infrastructure maintenance and rolling stock contracts put in place to enable privatisation to proceed, and
- these considerations operating in combination.

There is no reason to accept that privatisation, per se, will lead to a deterioration of service to the detriment of the consumer, or a loss of economic efficiency. Few would now argue that the privatised British Airways, British Steel, or British Telecom offer customers a poorer service than when these companies were nationalised. However, the proposition that privatisation leads to better customer service would not enjoy much support in relation to all privatised utilities. The quality of service offered to customers of the water utilities, for example, has come in for severe criticism in some parts of the country.

The poor service offered by some privatised utilities must lead to questions about whether this is an inherent defect of privatised monopoly, or at least a consequence of weaknesses in regulation. It is possible that the moderation of competition enjoyed by the franchised railway operators has resulted in some complacency on quality of service.

In questioning whether the vertical unbundling imposed upon the industry is intrinsically less efficient than a vertically integrated railway, Bill Bradshaw focuses on the operation of the infrastructure and train services. There is some evidence that managers in the railway industry, and those coming from without the industry, have taken time to learn how to operate the new contractual matrix. For example, some service quality problems appear to trace back to inexperience in managing staff and other resources.

Another reason for the rail industry's quality problems may lie in lack of robustness of the contractual matrix put in place to enable privatisation. Something of the order 70,000 contracts and leases were drawn up to knit the privatised railway industry together and it is not altogether surprising if some are now found to be wanting. For example, Railtrack in its 1998 Network Management Statement draws attention to the lack of alignment between itself and its customers in developing the railway network. This is a reflection of the contractual relationships put in place. There is also some evidence that key maintenance contracts have failed to provide the leverage necessary over suppliers. However, contracts can and will be rewritten, and new contracts, reflecting the experience of the first years of privatisation, should bring improvement.

The new structure imposed on the railway industry involves various transaction costs which directly stem from unbundling and the need for and elaborate contractual matrix. With greater experience, managers in the restructured industry may be expected to learn to write better contracts and to achieve higher quality. Assessing whether the efficiency gains unlocked by greater market testing outweigh the transaction costs inherent to the new structure will no doubt prove a fruitful topic for future academic research.

The Franchising Model in Rail and Bus

The rail franchising model adopted by the previous Government has been seen by some as a further source of inefficiency and poor performance. A number of reasons are advanced. First, it is argued that faced with an uncertain market for franchises, the Franchising Director chose to adopt less rigorous commitments on future franchise operators in terms of quality of service and particularly performance. Franchise agreements did not require any improvement above the standards previously obtained by British Rail, although the contracts provide incentives for doing so. Secondly, the majority of franchises are of relatively short duration, five to seven years with only a minority being for longer periods, in each case tied to significant rolling stock investment. Bill Bradshaw takes the view that the term of most franchises is too short to incentivise franchise operators to develop and improve their business. Thirdly, for a number of franchises the fare box represents only a limited incentive on the operator because costs heavily outweigh revenue. In these circumstances, it has been argued that cutting costs rather than improving service to the passenger and growing revenue, may be a much more effective way of achieving bottom line objectives. These issues are likely to form a key part of the agenda in any franchise renegotiation.

While the Government now looks to foster a closer relationship between bus operators and local authorities, the proposed extension of bus Quality Partnership arrangements still leaves bus operators in the 'driving seat'. Bus Quality Contracts would fundamentally change the relationship between bus operators and local authorities. Local authorities would in effect acquire property rights over the use of the road network for bus services which they would be free to sell to bus operators. This would change the structure of the bus industry, bringing a form of franchising to bus services. As with rail, the contractual relationships, including performance regimes and commitments to invest, the length of the contracts, and the scope for competition would be key issues.

Incentives to Invest

Bill Bradshaw draws attention to investment by Railtrack and that company's perceived unwillingness to invest in enhancement of the network. Railtrack has argued that the current charges regime gives it limited incentive to invest in the growth of the industry and highlights this issue in the 1998 Network Management Statement[16]. A key issue for the future, flagged by the Rail Regulator in his Second Consultation Document on the Periodic Review[17] is the need to give Railtrack adequate incentive to invest in improving the capacity and quality of the network.

In the case of the bus industry, the Government looks to Bus Quality Partnerships and possibly Quality Contracts to give a spur to investment and quality improvement. Achieving a robust contractual framework and firm commitments will be critical. An important issue will be whether relying on Quality Partnerships will be considered sufficient or whether Quality Contracts are needed to drive up quality.

The Role of Government

The White Paper recognises the need for a wider role for the public sector. It also sets out a framework in which the privatised bus and rail will be encouraged to play a larger role in meeting the nation's transport needs. It approaches the role of competition in bus and rail from the standpoint of the impact of competition on modal share.

Government is making additional public resources available for service improvement. Land use planning arrangements are to be changed so that greater account is taken of public transport provision. These are important initiatives but the ability of bus and rail to capture greater market share will continue to depend on offering value for money to the passenger and to the taxpayer. This will require high levels of cost effective investment, improving productivity and a coherent strategic planning framework.

A key question for the rail industry is how far the SRA will in future specify the outputs the rail industry is to deliver, or alternatively rely on market incentives to encourage growth. While the experience of the last few years may suggest that the former course is preferable, what are the implications for the rail industry in terms of efficiency and risk taking? For example, if the SRA proves to be over-prescriptive the consequence may be to dilute Railtrack's stewardship responsibilities for the network and repatriate risk back to the public sector. This is not necessarily in the passengers' or the taxpayers' interest. The issue will require careful consideration in formulating the remit and style of the SRA.

Similar issues arise for local bus services. A move to Quality Contracts would change the industry, transferring key responsibilities for the development of services to the public sector. Government and local authorities will need to determine the appropriate balance of responsibilities between private and public sector, and the implications in terms of risks and incentives.

Notes

1   White Paper, *A New Deal for Transport: Better for Everyone*, Cm 3950 July 1998.
2   *Railways Act* 1993, Chapter 43.
3   *Central Rail Users' Consultative Committee: Annual Report*, 1998.
4   The Central Rail Users' Consultative Committee, press release 3/98 16 March 1998
5   Para. 3.20 White Paper, *A New Deal for Transport: Better for Everyone*, Cm 3950 July 1998.
6   Para. 2.44 White Paper
7   Third Report of the Environment, Transport and Regional Affairs Select Committee, Session 1997-8. *The proposed Strategic Rail Authority and Railway Regulation.* House of Commons paper 286 - 1 March 1998.
8   Professor Bill Bradshaw. Paper given to the September 1998 Regulatory Policy Research Centre Seminar on *Integrated Transport Policy: Implications for Regulation and Competition.* Section 6.2. See pages 153-55 in current volume.
9   Professor Bill Bradshaw. Paper, Section 6.2. See pages 153-55 in current volume.
10   Professor Bill Bradshaw. Paper, Section 6.1. See pages 152-3 in current volume.
11   Office of the Rail Regulator, *Competition for Railway Passenger Services, A Policy Statement.* December 1994.
12   Office of the Rail Regulator, *New Service Opportunities for Passengers - Criteria for Evaluation,* June 1998.
13   Office of the Rail Regulator, Regulator's Foreword, *New Service Opportunities For Passengers - Criteria for Evaluation,* June 1998.
14   Para. 2.34, White Paper.
15   Para. 2.38, White Paper.

16 Railtrack.   *Network Management Statement 1998.*   Development of the
    Network page 4.
17 Office of the Rail Regulator. Section 5.5, *The Periodic Review of Railtrack's
    Access Charges: A Proposed Framework and Key Issues,* December 1997.

12 Essential Facilities and Transport Infrastructure in the EU

TREVOR SOAMES

Introduction

The question of access to essential facilities is a vexed one. Owners and operators of essential facilities consider it their legitimate right to determine who should gain access to the facility in question and on what terms. Conversely, businesses, whose activities in the related market are dependent on the provision of the essential facility, consider it their right to encroach upon the autonomy of the facility owner to enable them to compete.

Within the legislative framework of the European Community, the intervention of the competition authorities has played an important role in balancing these conflicting interests, though usually in favour of the market entrant. Application of the EC competition rules has permitted the authorities to exercise control over essential facility owners, to ensure that the facility is operated on terms which are fair, transparent and non-discriminatory and in accordance with what are considered to be proper procedures.

Investigations conducted by both DGIV and DGVII of the European Commission into the practices of owners or operators of transport infrastructures have proved the driving force behind the development of that body of EC case law which has increasingly become identified as that which relates to essential facilities. From the body of Commission decisions, a doctrine of increasing substance and definition is slowly emerging, on which the complainant of abusive behaviour on the part of the essential facility owner may rely to seek redress.

European Community Law

Among the fundamental tasks of the European Community is the establishment of a common market and an economic and monetary union

(Article 2 of the EC Treaty). Article 3(f) of the Treaty provides that the activities of the Community shall include a common policy in the sphere of transport. Article 3(g) of the Treaty provides that the activities of the Community shall include a system to ensure that competition in the common market is not distorted.

The European Commission plays an important role in ensuring the proper functioning and development of the common market. The Commission has a number of roles: it acts as guardian of the Treaties; serves as the executive arm of the Communities; initiates Community policies; and defends the Communities' interest in the Council. The Commission is divided into Directorates General (DGs) which are in charge of different fields of Community policy. The Directorates General of immediate interest to operators of transport infrastructures are DG IV which deals with competition policy and practice and DG VII which deals with transport. Although the application of the competition rules is the responsibility of DGIV (it will consult other interested services, such as DGVII when the case involves transport issues), decisions - and legislation - relevant to this discussion have also been adopted by DGVII in the exercise of its powers under specific transport legislation, rather than the general competition rules.

In general, EC competition law is concerned with the behaviour of undertakings which restrain competition. The articles of the EC Treaty which govern competition are as follows: Article 85 which prohibits undertakings, associations of undertakings from entering into anti-competitive agreements or concerted practices; Article 86 which prohibits one or more undertakings from abusively exploiting a dominant position; Article 90 which prohibits Member States from permitting public undertakings and undertakings which have been granted special or exclusive rights from acting in an anti-competitive manner; and Articles 92 and 93 which control the grant of aid which a Member State may give to undertakings where that aid would distort or threaten to distort competition.

In relation to the application of the competition rules to essential facilities, it is generally Article 86 which is the most important provision. Thus for the purposes of the present discussion, this paper will be focused primarily on the application of this article.

It should be noted that when the Competition Bill, presently in its final Parliamentary stages, becomes law and comes into force the EC case law described in this paper will become even more directly relevant to the UK, as the Chapter II prohibition of the Act is modelled on Article 86 and hence that form of control (and inter alia essential facility case law) will become applicable to behaviour in the UK. Should such abusive behaviour have insufficient effect on trade between Member States or be carried out by an undertaking which, though it may have a dominant position does not exercise that dominance in a substantial part of the common market, then Article 86 will not apply. By contrast, the Chapter II prohibition would apply due to the different criteria for its application and, as a result, the UK competition/regulatory authorities would act consistently with what is essentially a domestic Article 86 equivalent and use the EC case law to do so.

Article 86

Article 86 is an instrument for controlling the abusive exercise of monopoly power. Article 86 provides that:

> Any abuse by one or more undertakings of a dominant position within the common market or in a substantial part of it shall be prohibited as incompatible with the common market insofar as it may affect trade between Member States.

In contrast to Article 85, Article 86 applies to the unilateral conduct of a single undertaking (although it may, in certain circumstances, also be applied to a group of undertakings who enjoy a position of shared dominance). There is also no provision for an exemption. Any infringement of Article 86 is prohibited and national courts are entitled to apply it.

Dominance

An undertaking will be found to be dominant if it enjoys a position of economic strength which enables it to hinder the maintenance of effective competition on the relevant market by allowing it to behave to an appreciable extent independently of its competitors and customers and ultimately of consumers. A market share of over 50% almost certainly will be regarded as giving an undertaking a dominant position. In any event, in determining dominance the Commission examines the market share of all the other competitors on the market. The definition of the relevant market is of the utmost importance in determining the application of Article 86 and any undertaking seeking to allege dominance or defend itself against an allegation of abuse of Article 86 will pay a great deal of attention to how the relevant product and geographic markets are defined. In general terms, the undertaking alleging abuse will seek to define the market as narrowly as possible. The alleged abuser will seek to have the market defined as widely as possible and adduce cogent evidence of demand-side and supply-side substitutability in addition to pointing out the low barriers to entry so as to reduce its market share and deny allegations of dominance[1].

Article 86 may, in exceptional cases, also apply to companies which hold and abuse a 'collective dominant position'. In one case, *Italian Flat Glass*[2], the European Commission found that three Italian producers of flat glass occupied a collective dominant position since they could act together as an independent group without regard for the market behaviour of their competitors. In another case, *Cewal*[3], the European Commission found that the shipping companies operating a regular liner service between Zaire and Northern European ports, who were all members of the Associated Central West Africa Lines (Cewal) liner conference, held and abused a joint dominant position on the routes between those ports because of the high degree of co-operation between the shipping lines in the context of the conference[4]. The Commission also imposed massive fines on the parties to the Trans-Atlantic Conference Agreement (TACA) for allegedly abusing their joint dominant position in breach of Article 86[5].

Substantial Part of the Common Market

For Article 86 to apply, the undertaking concerned must not simply be dominant but must be so in a substantial part of the common market. In the *Sugar Cartel Case*[6] the Court of Justice ruled that the economic importance of the geographical market concerned must be considered to see if it is 'substantial'. The Commission has held that the economic significance of a part of the common market may be more important than the geographic area involved where the market is one for the provision of services at, or through, a facility that is part of the infrastructure of the transport or other industry in question. In a variety of cases, the Commission and the Court have demonstrated that in reality the threshold for demonstrating that dominance exists in a substantial part of the common market can be very low. For example, in the *Second Genoa*[7] case, Advocate General Van Gerven concluded that pilotage services in the port of Genoa constituted a substantial part of the common market due to the large quantities of freight going through Genoa, its market position in the context of all imports into and exports from Italy and the fact that pilotage is required for all ships calling at Genoa. Similarly in the *Port of Roscoff* case, the port of Roscoff was held to be dominant despite its small size and peripheral geographic location. The Commission held in its interim measures decision that in the narrowly-defined market of ferry services between Ireland and Brittany, Roscoff was the only port capable of providing adequate port facilities in France for ferry services between Brittany and Ireland.

Abuse

A dominant undertaking's behaviour will be considered abusive if it influences the structure of the market where, as a result of its behaviour, the degree of competition is weakened. The following are some examples of abusive behaviour: excessive pricing; predatory pricing; price discrimination; fidelity rebates and similar practices; refusal of access; grant of discriminatory access; abuse of intellectual property rights; tying clauses; and discrimination.

Predatory pricing has as its primary purpose the reduction of the target undertaking's profitability so that it has no choice but to exit the market or, in the case of predatory behaviour designed to deter entry in the first place, to ensure that the target undertaking knows that entry would be so expensive and painful that it is not worth investing in market entry at all. For a price to be predatory it must satisfy the two-tiered test set out by the European Court in the *AKZO*[8] case. The Court held that prices set below Average Variable Cost by which an undertaking seeks to eliminate a competitor must *per se* be considered to be abusive. Should the prices be above Average Various Cost but below Average Total Cost, then they will only be considered as predatory where there is evidence that they were set as part of a plan to eliminate a competitor. In practice, the AKZO test can be very difficult to apply and there have been some (so far unsuccessful) attempts to apply the concept to transport infrastructures[9].

Unfairly high pricing may exist where its purpose is designed to achieve for the dominant undertaking turnover and profits which it would not achieve in a more competitive environment. In the *United Brands*[10] case the ECJ looked at whether excessive prices had been charged. In doing so it asked:

> whether the difference between the costs actually incurred and the price actually charged is excessive, and, if the answer to that questions is in the affirmative, to consider whether a price has been charged which is either unfair in itself or as compared to the other competing products.

In practice, it has proved to be extremely difficult for the Commission to prove that a dominant undertaking has priced 'excessively'[11].

Fidelity rebates are special financial discounts granted by dominant firms which are designed, through the grant of a financial advantage, to prevent customers obtaining their supplies from competing producers.

The use of tying clauses by dominant undertakings also constitutes abusive behaviour. Tying clauses are clauses making the conclusion of contracts subject to acceptance by the other parties of supplementary obligations which, by their nature or according to their commercial usage, have no connection with the subject of such contracts.

A dominant firm should not discriminate between customers or suppliers unless there is a clear objective justification. The definition of what is an objective justification can be very difficult and is very much 'in the eye of the beholder'.

Many types of abusive behaviour can be regarded as discriminatory if rival competitors are harmed. Discrimination may be apparent where a dominant undertaking uses rebates and target discounts. Dominant firms should also avoid applying dissimilar conditions to equivalent transactions, as this too is an abuse under Article 86.

Refusal to supply can also constitute abuse of a dominant position. In the *Commercial Solvents*[12] case the Italian subsidiary of a US corporation was found to have abused its dominant position as the sole EC supplier of an essential raw material for the anti-tuberculosis drug when it stopped supplying the raw material to an Italian drug manufacturer, Zoja. Commercial Solvents had commenced manufacturing the finished product itself and wanted to halt competition. In the *British Midland/Aer Lingus*[13] case the termination by Aer Lingus of its interlining arrangements with British Midland and refusal to extend them to British Midland's new service between London (Heathrow) and Dublin was deemed to be abusive. Granting supply but on discriminatory terms may similarly be deemed abusive, for example in *Aeroports de Paris*[14], where third party suppliers were subject to a system of discriminatory licence fees.

Sanctions

There is no procedure for obtaining an exemption under Article 86; the prohibition on abusing market power is absolute. Once it has been established that a company has abused its dominant position, the European Commission may impose fines of up to 10 per cent of the aggregate annual world-wide turnover of the undertaking or group concerned[15]. Under the equivalent provisions of Chapter II under the UK Competition Bill, no exemptions can be granted and infringements will be subject to investigation and the imposition of fines.

The European Commission also has injunctive powers which enable it to order a company to stop abusing its dominant position at any time. An aggrieved competitor can make a complaint to the European Commission alleging abuse and if the European Commission concludes that the complainant is likely to suffer serious and irreparable damage, it can require the offending company to suspend its allegedly anti-competitive conduct.

Article 86 and Essential Facilities

Origins

The doctrine of essential facilities is simply an extension of the principles relating to refusal to supply, the subject of the refusal being the essential facility. It is correctly termed an extension of principles, since it has not been conceived as a self-standing set of tenets in its own right (although its proponents have sought to claim that is the case). Rather, it is the refinement of a nascent body of Commission jurisprudence in respect of bottleneck monopolies and leveraging. This was made abundantly clear from the cases and decisions cited by the Commission as authority for the legal propositions it first put forward in *the B&I v Holyhead (Holyhead I)* case, see recital 41 and footnote 3 to the decision. That said, with the Advocate General's Opinion in the case of *Oskar Bronner*[16], it is close to receiving full judicial recognition from the Court of Justice.

Principles

The essential facilities doctrine imposes a special responsibility on undertakings in a dominant position who either own or control access to an essential facility and places them under a duty to allow other entities access to that essential facility on a non-discriminatory basis. This responsibility is particularly important where the owner or operator of an essential facility (or another part of the same corporate group) itself *uses* that facility.

The Commission has defined an essential facility as:

> *a facility or infrastructure without access to which competitors cannot provide services to their customers*[17].

Refusal to provide access to a facility or the provision of access on discriminatory terms will constitute an abuse of a dominant position amounting to an infringement of Article 86. The abusive behaviour falls to be considered in two types of factual circumstances. First, where the owner and operator of the essential facility itself uses that facility and abuses its dominant control to favour its own activities in the related market (commonly referred to as 'leveraging') and secondly, where the owner and operator has no activities in the related market. In the first case, a presumption arises that the behaviour of the owner/user may well be motivated by a desire to disadvantage competitors in the related market. The dominant undertaking is thereby obliged to prove to the satisfaction of the Commission that its behaviour is objectively justified. In the second case, similar principles apply, but being in the position of pure operator/owner, no suspicion is aroused as to anti-competitive intent and thus the burden on the complainant of proving abusive behaviour is higher.

However wide this doctrine may appear, there would nevertheless seem to be limitations in the scope of its application. In *Oskar Bronner*, the Advocate General suggested that the doctrine may only apply where the dominant undertaking has a genuine stranglehold on the related market, making duplication of the facility impossible or extremely difficult. He added that where duplication of the facility alone is the barrier to entry (thereby echoing the dicta of the American doctrine) it must be such as 'to deter any prudent undertaking from entering the market'.

Comparison to US model

It is important to recognise that the essential facility doctrine has it origins, as with many other principles of EC competition law, in US antitrust law although, and as will be seen below, there are some significant differences in approach. Under US antitrust law, the doctrine of 'essential' facilities has primarily been derived from two 'bottleneck cases'. The first and most important of these cases was *United States v Terminal Railroad Association of St. Louis*[18], which involved a challenge to an association comprised of fourteen of the railroads serving St. Louis. The Association had acquired all the terminal facilities in St. Louis and all means of access to St. Louis. The Court held that the Association had violated the antitrust laws as when 'the inherent conditions are such as to prohibit any other reasonable means of entering the city, the combination of every such facility under the exclusive ownership and control of less than all of the companies under compulsion to use them violates' sections 1 and 2 of the Sherman Act.[19] The Court further required that the properties be made available to 'any other railroad not electing to become a joint owner, upon such just and reasonable terms and regulations as will....place very such company upon as nearly an equal a plan...as that occupied by the proprietary companies'.

The other case usually cited as providing authority for the doctrine of essential facilities is the judgement of the Supreme Court in *Otter Tail Power Co. v United States*[20]. Otter Tail was an integrated electric utility. Several municipalities in its service area attempted to enter the business of electric power distribution. Otter Tail attempted to prevent this entry in a number of ways, including refusing to sell or transmit wholesale power to the municipalities. The Court upheld the district court's finding that Otter Tail's practices constituted both an attempt to monopolise as well as actual monopolisation of 'retail distribution of electric power in its service area'. The Court considered that Otter Tail's conduct was unlawful because the use of monopoly power to destroy threatened competition was a violation of the 'attempt to monopolise' clause of section 2 of the Sherman Act.

In 1977, *Hecht v Pro-Football, Inc.*[21], became the first case to use the term 'essential facility doctrine'. In *Hecht*, a potential franchisee of the American Football League challenged a restrictive covenant in the Washington Redskin's stadium lease. The covenant prohibited leasing the stadium to any other professional football team. On appeal, the Court accepted the existence of the doctrine of essential facilities and stated that 'where facilities cannot practicably be duplicated by would-be competitors, those in possession of them must allow them to be shared on fair terms. It is illegal restraint of trade to foreclose the scarce facility'[22].

The Court also held that:

> To be 'essential' a facility need not be indispensable; it is sufficient if duplication of the facility would be economically unfeasible and if denial of its use inflicts a severe handicap on potential market entrants. Necessarily, this principle must be carefully delimited; the antitrust laws do not require that an essential facility be shared if such sharing would be impracticable or would inhibit the defendant's ability to serve its customers adequately.

The decision in *Hecht* was followed by the Seventh Circuit's decision in *MCI Communications v American Telephone & Telegraph Co*[23]. In that case, the Seventh Circuit considered a challenge by MCI to a number of AT&T's practices, including AT&T's refusal to grant MCI access to local Bell facilities. The Court of Appeals sustained a jury finding that AT&T's action violated section 2 of the Sherman Act. The Court cited the essential facility doctrine as the basis of its decision and stated that there are four elements necessary to establish liability under the essential facilities doctrine, namely:

- 'control of the essential facility by a monopolist;
- a competitor's inability practically or reasonably to duplicate an essential facility;
- the denial of the use of the facility to a competitor; and
- the feasibility of providing the facility'.

The Seventh Circuit also imposed liability under the essential facilities doctrine in the important case of *Fishman v Estate of Wirtz*[24]. This case involved competition to acquire the Chicago Bulls basketball team. One of the competing groups alleged that another competitor used a variety of unfair means to gain an advantage in negotiations, including the refusal to lease the Chicago Stadium which was owned by a member of the other group. The Court affirmed a lower court finding that the Chicago Stadium constituted an essential facility and that the refusal to lease the stadium violated the Sherman Act. The Court considered that the Chicago Stadium was an essential facility because it was substantially superior to any existing alternative and dismissed the fact that a new stadium could be built, and actually was built, on the grounds that 'a potential market entrant should not be forced simultaneously to enter a second market, with its own large capital requirements'[25].

Another notable recent case was *Consolidated Gas Co. of Florida v City Gas of Florida Inc*[26]. That case involved a dispute between two natural gas distributors. City Gas was a large distributor in Southern Florida; Consolidated Gas was a much smaller distributor in much the same area. Consolidated Gas wished to serve a group of City Gas's customers, but could not economically serve without access to the large company's pipeline. City Gas, however, would only permit access to its pipeline at an unreasonably high price. Holding that City Gas's unreasonable offer was tantamount to a refusal to deal, the Court held that City Gas was in violation of section 2 of the Sherman Act under the essential facilities doctrine.

As will be seen below, the position under EC competition law is somewhat different to that of the USA: particularly as regards its limited application under US law where granting access to the essential facility is not considered to be 'feasible'. However, the first and most important point of similarity between the essential facilities concept under EC law, as compared to that under US antitrust law, is that under the latter the essential facility doctrine is only applicable where the facility is controlled by a 'monopolist'. In EC jurisprudence it is, of course, an essential prerequisite for the application of Article 86 to the control of essential facilities, that the owner/operator of the essential facility occupies a dominant position in a substantial part of the common market in respect of that facility or the services which use that facility (or both). In circumstances where the owner/operator of the essential facility does not occupy a dominant position within the meaning of Article 86 then, subject to national competition/regulatory law, it does not have the same obligations to provide non-discriminatory access as dominant undertakings do.

Nonetheless, despite the apparent clarity of the written word, the American doctrine has not escaped criticism for the lack of clear direction in its practical application[27].

Application

So far, the Commission has found a variety of owners/operators of essential facilities to be in breach of Article 86 in respect of, for example, access to port facilities at Holyhead, Roscoff, Elsinore and Rødby; access to pilotage services and freight handling at the port of Genoa; access to a computer reservation system; access to refuelling at Milan's Malpensa Airport; access to ramp-handling at Frankfurt airport; access to gas pipelines; access to telecommunications; access to Orly and Linate airports. For the purposes of the present discussion, however, examination of the application of this doctrine will focus on two key infrastructures of the maritime and aviation industries - ports and airports.

Application of Article 86 and the Doctrine of Essential Facilities and the Transport Sector

Application to Airport Facilities

Access to airport facilities is frequently the subject of infringement proceedings under Article 86. As explained above, an undertaking will be in breach of Article 86 if it has a dominant position in a substantial part of the common market and abusively exploits that position in such a way that it affects trade between Member States. In order for the essential facilities doctrine to apply, all the elements of Article 86 must be present.

Dominance and Access to Airport Facilities

In determining whether the operator of an airport facility is in a dominant position it is necessary to identify the relevant market. The operator of an airport facility seeking to dispute an allegation of abuse of a dominant position will seek to define the relevant product and geographic markets as widely as possible so as to reduce the possibility that it may seem to be dominant. It will also argue that the facilities in question are not 'essential' and that, even if they are, access to them on the terms requested is not feasible.

In a number of aviation cases the Commission has defined the market quite narrowly. In the *British Midland/Aer Lingus* case the Commission considered the relevant market to be that for the provision and sale of air transport between Dublin and London Heathrow (and not the provision and sale of all transport including surface transport or even the market for air transport between all London airports and Dublin).

Airports are often natural monopolies except to the extent that competition is provided by a neighbouring airport. In assessing whether an airport or airport facility is truly dominant, it is necessary to examine whether it is substitutable. This involves an assessment of whether there are alternatives to the airport or facility in question which are realistic substitutes. In addition, it may be possible to argue that there are alternatives to the airport or its facilities where surface transport is a realistic alternative to air transport, and ascertain whether the route can easily be substituted by other routes between the destinations in question. The ease of travel and duration of transportation will also be a factor.

An airport could seek to assert that it could be substituted for other airports or facilities in the same vicinity and even with airports in neighbouring countries. However, this may be difficult to argue successfully. In the *Zaventem* decision[28], the Commission found that there were no genuine alternatives offering the same advantages as Brussels Airport as it was of limited substitutability with other available routes for short and medium-haul services within the Community operating from or to the Brussels catchment area and faces only minor competition from them. In the Commission's decision in the *British Midland/Aer Lingus* case, it found that London Heathrow was the relevant airport in the London area and other London airports were not considered to be substitutable as business travellers traditionally prefer Heathrow because of the frequency of flights and the range of connections available (it stated that, in any event, even if all London airports were included in the definition, Aer Lingus would still be dominant). It was concluded in *Aéroports de Paris,* that the high volume of national and international connections at Orly and CDG combined with the lack of comparable airports in the vicinity left suppliers of ground-handling services with 'no other choice than to use the Orly and CDG airports'. Similarly in *Frankfurt Flughafen*[29], the Commission found that no other airports in the near vicinity of Frankfurt could be considered substitutable for Frankfurt airport for the provision of point-to-point air transport services to and from Frankfurt and the surrounding region.

Where an undertaking controls a number of airports operating in the same geographic market it could be easier to prove that the owner of the airports in question is in a dominant position in respect of the provision of airport facilities in the geographic markets. A case in point is again the *Aéroports de Paris* decision, where ADP, as the body responsible for the management of Paris airports and thus essential to the proper operation of national and international air transport services to and from the Paris region, was held to occupy a dominant decision.

If a facility can be described as essential for competitors to gain access to a given market, then the operator of that facility is *prima facie* in a dominant position in respect of that facility.

An Airport Facility as a Substantial Part of the Common Market

The Commission has had no difficulty in proving that airports or airport facilities constitute a 'substantial part of the common market'. The Commission has considered a wide variety of undertakings to hold a dominant position in a substantial part of the common market in respect of their control of essential facilities including Brussels Airport (*Zaventem*);

the Belgian market for the provision of computerised reservation services (*London European - Sabena*)[30]; Paris Orly and CDG (*Aéroports de Paris*); access to refuelling at Milan's Malpensa Airport (*Disma*)[31]; access to ramp-handling at Frankfurt (*Flughafen Frankfurt/Main AG*). In the last case, basing its reasoning on the Court's judgement in *Port of Genoa*[32], Frankfurt airport was considered by the Commission to constitute a substantial part of the common market, being the largest international airport in Germany and accounting for the largest volume of traffic in terms of passengers, freight or commercial aircraft movements.

Airport Facility as Essential Facility

An essential facility is a facility or infrastructure without access to which competitors cannot provide services to their customers. There is almost certainly a large number of airport services which can be regarded as essential to the provision of an air service. These would include access to slots, access to the ramp, access to baggage handling, access to gates and lounge space, access to refuelling and maintenance facilities, access to catering facilities, air traffic control, perhaps even access to a hangar or space to store aircraft when not in use.

The Commission is frequently willing to intervene to ensure that entities wishing to gain access to an airport facility gain that access wherever possible and on non-discriminatory terms[33]. In addition, when the Commission monitors acquisitions, it takes note of dominance and the question of access to essential facilities. For example, in the planned acquisition by Lufthansa of Interflug in 1990[34], the Commission sought to ensure that facilities which are essential to all airport users were not controlled by a company in a dominant position.

Abuse by the Operator of an Airport Facility

The operator of an airport facility which is in a dominant position in a substantial part of the common market must take care to avoid acting in such a way as to influence the structure of the market where, as a result of its behaviour, the degree of competition is weakened. The operator should ensure that it acts in a fair and non-discriminatory manner at all times and that it carries out all its procedures according to the rules and makes all its decisions according to relevant, objective, transparent and non-discriminatory criteria.

The range of possible abusive behaviour by an operator of an airport facility is very wide. However, the most likely abuses by the operator are (i) refusing the competitor access to the facility which it controls or (ii) granting the competitor access on less favourable terms than those which the operator itself enjoys or that are enjoyed by other competitors. This might involve limited access, access at unfairly high prices, or on other less favourable and discriminatory terms. In the absence of objective justification, such abuses will be caught by the provisions of Article 86.

The most recent case before the Commission, *Aéroports de Paris*, concerned the application of dissimilar conditions to suppliers of ground-handling services at Paris airports. Following a complaint by Alpha Flight Services Sarl ('AFS'), a supplier of catering services at Paris-Orly, Aéroports de Paris ('ADP'), in its capacity as manager of the Paris airports, was found to have infringed Article 86 by imposing discriminatory fees on suppliers of ground-handling services.

In granting licences to providers of ground-handling services, ADP charged both a fee for the occupation of the site and a commercial fee based on turnover. From business correspondence with its customers, AFS learned that the level of its commercial fee was far in excess of that charged to a competitor. Some of the customers' letters even cited the commercial fees imposed by ADP as a reason for AFS charging high prices. The discriminatory effect of the higher fees substantially impaired AFS's ability to compete: it had either to align its prices with those of its competitors (and in so doing sacrifice a part of its profit equal to the difference in fees) or simply lose the customers. In coming to its decision, the Commission also had regard to the discrepancy between external fees (for suppliers of third party services) and internal fees (for self-handling operators, some of which were also licensed to provide third party services), where, in exchange for identical airport management services, the low rate of the latter effectively meant subsidisation by the former.

In its reasoning, the Commission made the distinction between the market for airport management services and the market for suppliers of ground-handling services or self-handling users and concluded that anticompetitive behaviour on the first market would directly affect market conditions on the second. The geographic market was considered to be the Paris airports of Orly and CDG. Bearing in mind their status as European hubs and the geographical remoteness of other hubs capable of offering equivalent services, there was no possibility of substitution. As the sole authority empowered with determining the conditions under which ground-handlers should operate, ADP was clearly dominant as suppliers were placed 'in a position of considerable dependency on ADP'. To establish whether ADP was abusing its dominance, it fell to the Commission to determine whether the differences in fees created dissimilar conditions for equivalent transactions. It concluded that:

> In the present case, ADP does not apply any commercial fee system which fixes in advance the rates for the commercial fee based on turnover. In exchange for the airport management services provided by ADP, the commercial fees vary individually according to supplier or user engaged in the same ground-handling activity. The fee charged within the same airport varies from one supplier to another and from one user to another, thus having an appreciable effect on the cost of the services concerned and on the structure of costs borne by the carriers. Such discrimination has anti-competitive effects on the market for air transport services.

A more severe form of abuse is apparent where the operator itself uses the essential facility. In a number of cases, operators have abused their dominant control of the essential facility to protect their own position in the related secondary market, either by offering the competitor access to the facility on less favourable terms than the operator itself enjoys or refusing the competitor access altogether.

The abusive conduct in *London European-Sabena* consisted of Sabena's refusal to grant London European access to its Saphir computer reservation system without a tie-in. Access to the Saphir system was essential to any company wishing to compete on the Belgium air transport market. Concerned with the low level of London European's fares, which were half those of Sabena's, Sabena's motivation was clearly to place indirect pressure on London European to raise its fares. The unlawful manner in which it attempted to distort market conditions to protect its own interests, by refusing access to its reservation system, was deemed to be an abuse of its dominance.

In the *Disma* case, the Commission found that access to a fixed aircraft-refuelling installation at Milan's Malpensa airport was essential to oil companies wishing to supply their customers. The Commission found that the proposed provision, preventing companies who were not members of the consortium which operated the refuelling installation from gaining access to the facility on non-discriminatory terms (the members of the consortium envisaged imposing significantly higher duties), would be considered abusive conduct, as would be the proposed clause making it difficult for new members to join the consortium.

The Commission has also sought to apply the provisions of Article 86 to providers of air transport and air transport related services (such as ground handling) where the undertaking concerned enjoys a monopoly conferred by the Member State in the provision of that service. For example, the ground handling monopolies in Spain and Greece enjoyed by Iberia and Olympic respectively have been investigated by the Commission under the combined provisions of Articles 86 and 90, as have the monopolies held by the airport operators of Milan and Frankfurt[35]. The authority to apply Article 86 and related provisions in the circumstances of State-derived monopolies, was provided by the Court of Justice in the case of *Telemarketing*[36]:

> Article 86 applies to an undertaking holding a dominant position on a particular market, even where that position is due not to the activity of the undertaking itself, but to the fact that by reason of provisions laid down by law there can be no competition or only very limited competition on that market.

With regard to the monopoly of the Frankfurt airport operator, the Commission has most recently adopted an Article 86 decision. The Commission started its investigation following a complaint, which was lodged on 20 July 1993 by KLM, Air France and British Airways. The facts of the case are simple. FAG, the company which owns Frankfurt Airport, has the monopoly for the operation of the airport. In its capacity of airport operator, it refused to allow self handling or third party handling for ramp-side ground-handling activities[37] (apart from fuelling and catering).

In its decision, the Commission concluded that the refusal of FAG to allow competition for the provision of ramp-side ground-handling services constituted an abuse of a dominant position within the meaning of Article 86. The Commission's reasoning can be summarised as follows.

The case concerned two closely linked but nevertheless distinct service markets, namely the market for the provision of airport facilities for the landing and take-off of aircraft in the Frankfurt area and the market for the provision of ramp-side ground-handling services within Frankfurt airport.

With regard to the provision of airport facilities, the relevant geographic market was considered to be Frankfurt airport, in view of its size and status as a major European hub. Given that FAG had a monopoly for the operation of the airport, it benefited from a dominant position in the relevant market for the provision of airport facilities.

With regard to the provision of ramp-side ground-handling services, the Commission first pointed out that although the provision of ground-handling services are complementary to the operation of airport facilities, they are two distinct markets. Indeed, the ramp-side ground-handling services may be (but do not necessarily have to be) provided by the airport operator. As the sole provider of ramp-side ground-handling services, FAG had a dominant position in this market.

After having established that FAG enjoyed a dominant position, the Commission examined whether FAG's refusal constituted an abuse of that position. It considered that that was the case, given that:

> FAG has made use of its power as exclusive provider of airport facilities to deny potential competitors (in the market for the provision of ramp-handling services) access to the ramp. FAG has thereby prevented potential suppliers of ramp-handling services from entering the market for the provision of those services. This applies both with regard to airlines and to independent suppliers. FAG has thereby monopolised the market for the provision of ramp-handling services. In deciding to retain for itself the market for ramp-handling services at Frankfurt airport, FAG has extended its dominant position on the market for the provision of airport landing and take-off facilities to the neighbouring but separate market for ramp-handling services.
>
> FAG has furthermore made use of its power as exclusive provider of airport facilities to deny airlines the right to self-handle. FAG has thereby obliged the users of its airport facilities also to purchase from it the ramp-handling services that they need.

The Commission then examined whether there were any objective reasons (such as limited parking space and traffic density or public service obligations) which excused FAG's behaviour, but concluded that there were none. Of particular interest was FAG's attempt to justify its behaviour on the ground of historical developments. It maintained that its monopoly had developed from the immediate post-war period, a time where responsibility for provision of ramp-handling services fell to the airport alone. Potential competitors only became interested in gaining access to the ramp-handling market once it transpired that this market offered attractive economic opportunities at little risk. The Commission agreed that given the lack of would-be suppliers, FAG's monopoly on the ramp-handling services was the result of market conditions and was therefore not illegal. However, infringement of Article 86 within the meaning of the *Telemarketing* judgement, arose as soon as FAG's monopoly position was maintained by a refusal on its part to authorise self-handling or third-party handling. At that point, FAG's monopoly was no longer the result of market conditions, but the conscious decision on the part of FAG, in the dominant position for the market of the provision of airport facilities for the take-off and landing of aircraft, to reserve for itself the neighbouring market for ramp-handling services.

Finally, the Commission examined whether FAG's refusal was justified by Germany's decision of 29 September 1997 to exempt FAG, in accordance with Article 9(1) of the Ground-handling Directive[38], for a limited period of time from the liberalisation requirements of the Directive. The Commission decided that this was not the case, given that: (a) the decision of the German Government could only become effective in the absence of Commission opposition under Article 9(5) of the Directive; and (b) in any case, the provisions of the Directive are without prejudice to the rules laid down in the Treaty, including *inter alia*, EU competition rules.

In point of fact, on the same day the Commission adopted a decision[39] pursuant to Article 9(5) of the Ground-handling Directive requiring Germany to amend substantially the scope of its decision granting an exemption to FAG. The effect of the Commission decision was to limit the derogation in time, until December 2000, and in area of application, the eastern cul-de-sac of Terminal 1 only.

Notwithstanding the adoption of these decisions, FAG nevertheless attempted to preserve the substance of its old monopoly, by tying in a large number of airlines to long-term contracts of three to ten years' duration. The Commission, however, was quick to react and immediately informed the airport of the illegality of this practice. Confronted with the possibility of further infringement proceedings, FAG agreed to change its policy on long-term contracts, by allowing its customers to terminate contracts without penalty subject to satisfying a six months' notice provision[40].

All the above cases illustrate the Commission's efforts to strike down barriers to competition at national airports, a task made all the more challenging by the fact that airports are traditionally the province of state-derived monopolies.

Although the reasoning in these cases attests to the operation of the essential facilities principles in practice, the more explicit wording of the doctrine is absent. A more clearly-articulated version of the doctrine has developed in the case-law relating to access to sea ports.

Application to Sea Ports

It has been said above that the doctrine of essential facilities operates provided all the elements of Article 86 are present[41].

In respect of access to ports, the two Holyhead cases, relating to the operation of that port by the Stena Sealink group, have contributed a great deal to the development of this doctrine[42].

The port of Holyhead is in North West Wales and serves the short-sea route between Britain and Ireland. The nature of the market is such that the Commission has concluded that the short-sea market between Holyhead and Ireland constitute a separate market from other, potentially competing, routes to the north and to the south (whose journey times are said to be considerably longer than that between Holyhead and Ireland). The Commission concluded in its two decisions that the relevant market was that for the provision of port facilities for passenger and vehicle ferry services on the central corridor routes between the United Kingdom and Ireland and that, after having noted that there were no reasonably substitutable ports (including Liverpool), took the view that the port of Holyhead was the only port serving that market on the British side, giving Stena Sealink, in its capacity as port authority, a dominant position. The Commission concluded that the port of Holyhead constituted an essential facility because no other ports could be used without increasing substantially the journey time and that it was not feasible for either B&I or Sea Containers to build a new port (the Commission considering that the latter course would have been neither economically nor physically realistic). Furthermore, the Commission concluded that not only was the port of Holyhead in a dominant position but also that dominance was exercised in a substantial part of the common market; the port itself constituted that 'substantial part' as it provided one of the main links connecting Great Britain with the capital city of Ireland and the most popular ferry route between Ireland and Great Britain.

The port of Holyhead was owned and operated by the Stena Sealink group which also operated its own roll-on roll-off ferry service on the route between Holyhead and Ireland. In early 1992, a formal complaint was submitted to the Commission alleging that because Stena Sealink's ferry operating subsidiary had altered the sailing times of its ferries from the port and that its associated sister company, Sealink Harbours Ltd, in its capacity as port owner and operator, had allowed it to do so, the operations of B&I at the port were improperly disadvantaged.

The case is complicated and the decision that the Commission ultimately adopted hardly makes the position very much clearer. The essential problem was that the port of Holyhead is shaped rather like a bottle and that the berth used by the Stena Sealink ferries (Station Berth) was at the base of that bottle. In order for Stena Sealink vessels to pass to or from the open sea, they had to pass through the channel leading from the inner harbour to the open sea, which is narrow, rather like the neck of a bottle. The Admiralty pier berth which the B&I ferry used at that time was in that channel and every time a Stena Sealink ferry passed between the open sea and the inner harbour, the resulting wash sometimes caused the B&I ferry to move in its moorings, necessitating the disconnection of the linkspan (i.e. the ramp) between it and the land. This disconnection had to be undertaken for safety reasons primarily because the B&I vessel, being extremely old and improperly adapted to the port did not fit tightly against the linkspan. Before Stena Sealink changed its ferry operating schedule, during each of the two times per day that a B&I ferry was berthed at the port, a Stena Sealink ferry would pass once through the narrow channel, into or from the inner harbour, resulting in B&I disconnecting the linkspan for a short period of time.

For commercial reasons Stena Sealink decided to change its ferry operating schedule but this had the result that a Stena Sealink ferry would pass through the channel between the inner harbour and the sea and, therefore, the B&I vessel berthed at Admiralty Pier twice, rather than once, during each time that B&I vessel was berthed at Holyhead. B&I claimed that the resulting disruption and inconvenience to its services was such as to cause it severe commercial harm and complained to the European Commission that Stena Sealink in its capacity as port operator had abused its dominant position, contrary to Article 86 of the EC Treaty by allowing its associated ferry company to alter its schedules at Holyhead in such a manner as to cause disadvantage to its competitor, B&I.

The Commission applied a version of the US anti-trust doctrine of essential facilities to the Holyhead case. In its key statement of principle at recital 41 of the decision, the Commission stated:

> A dominant undertaking which both owns or controls and itself uses an essential facility, i.e. facility or infrastructure without access to which competitors cannot provide services to their customers, and which refuses its competitors access to that facility or grant access to competitors only on terms less favourable than those which it gives its own services, thereby placing the competitors at a competitive disadvantage, infringes Article 86, if the other conditions of that Article are met. A company in a

dominant position may not discriminate in favour of its own activities in a related market ... The owner of an essential facility which uses its power in one market in order to strengthen its position in another related market, in particular, by granting its competitors access to that related market on less favourable terms than those of its own services, infringes Article 86 where a competitive disadvantage is imposed upon its competitor with objective justification[43].

Furthermore, the Commission held that the owner of the essential facility, which also uses the essential facility:

... may not impose a competitive disadvantage on its competitor, also a user of the essential facility, by altering its own schedule to the detriment of the competitor's service, where, as in this case, the construction or the features of the facility are such that it is not possible to alter one competitor's service in the way chosen without harming the other's. Specifically, where, as in this case, the competitor is already subject to a certain level of disruption from the dominant undertaking's activities, there is a duty on the dominant undertaking not to take any action which will result in further disruption. That is so even if the latter's actions make, or are primarily intended to make its operations more efficient. Subject to any objective elements outside its control, such an undertaking is under a duty not to impose a competitive disadvantage upon its competitor in the use of the shared facility without objective justification, as seemed to be accepted by [Stena Sealink] in 1989[44].

The Commission held that Stena Sealink, in its capacity as port operator, had failed to comply with its duties by allowing its ferry operating subsidiary to change its schedule in a manner which the Commission concluded imposed a competitive disadvantage on B&I (because of the alleged additional disruption caused by the second disconnection of the linkspan whilst the B&I vessel was berthed at Holyhead). The Commission adopted a stern test:

Even if Sealink as ferry operator altered its schedule because it considered that it was in its commercial interests to do so, and Sealink as harbour authority allowed this, the result is still a prima facie abuse of a dominant position by Sealink, since it is to improve that undertaking's position at the expense of its competitor, B&I. A dominant company may improve its service but if that improvement will necessarily harm its competitor then its own commercial interests are not enough for the purposes of Article 86 to justify the resulting harm to the competitive situation of its competitor.

The Commission concluded that there was a *prima facie* case of infringement of Article 86 and that, in the circumstances of the case, interim measures were justified. In its decision the Commission ordered Stena Sealink to alter its ferry schedule so as to prevent the B&I ferry being disturbed more than once whilst berthed at Holyhead at each of its two slot times (one in the early morning and one in the afternoon): so restoring the *status quo* pending the Commission's final decision. Stena Sealink made an application to the Court of First Instance in Luxembourg for interim measures to suspend the effect of the Commission decision as well as an application for annulment. After a hearing before the Court, at the suggestion of the President, a compromise solution was reached whereby Stena Sealink altered its schedule so that the B&I ferry was not disturbed more than once at its berth during just one of its slots, whilst during the other slot it could be disturbed twice. The compromise resulted in the withdrawal of the applications for interim measures and annulment.

At the end of 1993, the Commission adopted another decision in respect of the port of Holyhead, this time relating to an application for interim measures by Sea Containers, the former owner of the Stena Sealink group and the owner of the Hoverspeed group, which operates hovercraft and fast ferry (SeaCat) services. In this case, Sea Containers submitted a wide-ranging complaint to the European Commission primarily alleging that the planned redevelopment for the port of Holyhead would have the effect of reducing the number of available berths at the port and thereby reducing its capacity. The Commission was asked to adopt interim measures to halt or amend the re-development plan. Furthermore, Sea Containers alleged that Stena Sealink had refused Sea Containers' access to the port for the operation of its fast ferry SeaCat service.

In a factually complex case, the Commission ultimately decided to reject the application for interim measures on the redevelopment issue and also concluded that interim measures could not be adopted on the access issue. This was on the basis that it was reasonable for Stena Sealink to offer Sea Containers slots at the existing berthing facilities at the port and to refuse Sea Containers' demand that it be allowed to build a new and dedicated berth for its fast ferry operation on that part of the port which had been identified for redevelopment. It was Stena Sealink's contention that if Sea Containers' request was granted, it would block or significantly inhibit the re-development of the port to the disadvantage of all users and the local community. It also argued that it had never refused access to the port of Holyhead and that, indeed, Sea Containers had accepted an offer of slots made for 1993 operations but had later withdrawn its acceptance.

In its decision, the Commission rejected the application for interim measures on the basis that although it considered there to be a *prima facie* case of infringement of Article 86 on the access issue (in relation to 1993 operation), no serious and irreparable harm would be caused to Sea Containers in the absence of interim measures as Stena Sealink had offered for 1994 operations the exact slot times that Sea Containers had originally requested (and stated to be essential for its operations) at existing port facilities.

If one reads the case carefully, one can see that the whole case revolved around the issue of four requested slot times per day[45]. Of those times, three were never in dispute but the fourth, which was at the busiest period of the day, with a proposed departure time of 14.00, was considered by Stena Sealink (and supported by B&I) as being likely to cause significant disruption at the port to the disadvantage of existing users and passengers. On that basis, Stena Sealink refused the request of Sea Containers to operate at that fourth identified slot time and instead suggested an alternative which was different by about half an hour. It was around this half an hour that the whole of the case ultimately revolved around! The disputed half hour was only conceded by Stena Sealink after the oral hearing of the case before the Commission and following remedial measures designed to alleviate port congestion.

The moral of the story is that ports (or any other essential facility operator) which occupy dominant positions within a substantial part of the European Union may be expected to have their activities very closely reviewed. This was made clear in those recitals of the Commission's decision which criticised Stena Sealink's conduct of the negotiations regarding Sea Containers' access to the port. The Commission expressed concern at the alleged delays and difficulties put in the way of Sea Containers' use of the port and Stena Sealink's alleged failure to negotiate:

> Sealink did not conduct its negotiations with Sea Containers by proposing or seeking solutions to the problems it was raising and that its rejection of all of [Sea Containers'] proposals without making any counter offer or attempting to negotiate was not consistent with the obligations on an undertaking which enjoys a dominant position in relation to an essential facility[46].

The Commission further criticised Stena Sealink for having been 'entirely negative' regarding Sea Container's proposals and whose approach (in the eyes of the Commission) 'consisted of raising difficulties'. Furthermore, the Commission considered that Stena Sealink should have set up procedures for 'dealing with its responsibilities as harbour operator to ensure that it carried out its duties to other operators'. In particular, the Commission considered that Stena Sealink did not consult sufficiently with Sea Containers and B&I about timetables and port development plans and that it should have considered reconciling the 'interests of existing and proposed users of the port' by making 'modest changes in the allocated slot times or in any plans for the development of the harbour'[47]. The limitation on the doctrine of essential facilities, under US antitrust law, where it is not feasible to grant access to the facility/infrastructure without adversely affecting the quality of service provided to existing users was, apparently, not recognised by the Commission or, perhaps, was applied with a generous interpretation of what is 'feasible': see, for example, *Hecht* and *Southern Pacific Communications v AT&T*[48].

The Commission took the view in the Holyhead cases that a port occupying a dominant position has certain obligations with which it must comply (*e.g.* non-discrimination) but indicated that a port which occupies a dominant position and which is operated by an undertaking which itself operates competing services from that port owes duties which may be significantly greater. For example, the Commission indicated that Stena Sealink should have contemplated requiring its ferry operating company to change its (or B&I's) schedule to accommodate Sea Containers' preferred slot times in a manner whereby congestion problems were minimised (even to the commercial disadvantage of existing users) or to alter the port redevelopment plans themselves. In other words, the Commission expected the port operator to go to great lengths to find 'feasible' means of access.

It is essential to note that where the operations of a port, which constitutes an essential facility (and, for obvious reasons, not all ports will fall into this category if, for example, there are alternatives which are realistic substitutes) and where the undertaking controlling that port's operations is able to confer an advantage upon that undertaking's group in a different, but related market, the burden of proof is placed upon the port operating undertaking to demonstrate that, where port operations impact negatively upon other users (or potential users), it has acted for objectively justifiable reasons. Put another way, in the absence of clear evidence to the contrary, decisions of a dominant undertaking *qua* port operator which have an adverse effect on actual or potential competitors, may be presumed by the Commission to have been taken for anti-competitive and abusive purposes. It might be a mistake, therefore, to read the statements of principle made in the *B&I* and *Sea Containers* cases as being applicable to all dominant ports because those particular cases related to situations where the dominance of the port was exercised within the context of the controlling undertaking's group being involved in other, but related, activities using that port (i.e.. ferry operations) and, as a result, Stena Sealink was seen by the Commission as being tempted to discriminate against competitors in those other activities.

In another critical port decision but this time under Article 90, the Commission built upon the essential facilities doctrine of Article 86. In the *Port of Rødby* case[49], the Commission held that the refusal by the Danish Government to allow a company to build a new port in the immediate vicinity of the port of Rødby or to operate from the existing port facilities at Rødby was in breach of Article 90(1) of the EC Treaty read in conjunction with Article 86. The complainants in this case were Europort A/S, a Danish company, and Scan-Port GmbH, a German company, both of which were subsidiaries of the Stena group which wished to operate ferry services between Denmark and Germany. By two separate requests, Europort had applied to the Danish Transport Minister for permission to operate from the port of Rødby and to build a private commercial port in its vicinity. Both requests were refused.

It was significant in this case that utilisation of the port terminals required the authorisation of the Danish Transport Minister acting on a proposal from the DSB, a public undertaking which owned the port of Rødby and which was responsible for its management. DSB had an exclusive right to organise rail traffic in Denmark and it also operated ferry services between Denmark and neighbouring countries, although it did not have any exclusive rights to operate such services. It was also significant that the route between Rødby in Denmark and Puttgarden in Germany was operated jointly by DSB and Deutsche Bundesbahn, the German public railway undertaking. No other companies provided ferry services on this sea route.

The Commission relied on two provisions of the EC Treaty to hold that the refusals infringed EC competition law. Article 90(1) provides that in their dealings with public undertakings and undertakings granted special or exclusive rights, Member States must neither enact nor maintain in force any measure contrary to the EC Treaty rules. DSB was a public undertaking and the two refusals by the Danish Transport Minister constituted state measures within the meaning of Article 90(1). However, there can be no breach of Article 90(1) unless another Treaty rule has been infringed. In this case, the Commission found an infringement of Article 86, which prohibits any abuse by one or more undertakings of a dominant position within the EC or in a substantial part of it, insofar as it may affect trade between Member States.

In applying Article 86, the Commission defined the relevant market as the market for the organisation of port services in Denmark for ferry services operating on the Rødby-Puttgarden route (for both passengers and vehicles). As regards sea transport between Eastern Denmark on the one hand and Germany and Western Europe on the other, there was no real alternative with the same advantages as those offered by the port of Rødby and air transport was far more expensive and limited to small items of freight and passengers without cars. It was noted also that 70.8% of travellers and 87.9% of lorries crossing by sea between Denmark and Germany used the Rødby-Puttgarden route. The market for maritime transport services between Rødby and Puttgarden was found to be a neighbouring, but separate, market that might be affected by the behaviour of an undertaking on the relevant market.

The Commission found that, by virtue of its exclusive right as port authority, DSB held a dominant position on the relevant market. In addition, DSB and Deutsche Bundesbahn held a joint dominant position on the Rødby-Puttgarden route as they were the only companies operating on it. This dominant position was protected by state measures relating to the relevant market.

In determining whether there had been an abuse of the dominant position on the relevant market, the Commission stated:

> Thus an undertaking that owns or manages and uses itself an essential facility, i.e. a facility or infrastructure without which its competitors are unable to offer their services to customers and refuses to grant them access to such facility is abusing its dominant position[50].

Consequently, the owner/operator of an essential port facility, which provides shipping services from it, may not 'without objective justification' refuse access to a competing shipowner. Although not specifically cited in the Rødby decision, the Commission was drawing on its previous formulations of the essential facilities doctrine in the access to Holyhead decisions (*B&I/Sealink, Sea Containers v. Stena Sealink* reviewed above).

The refusal to allow Europort to operate from Rødby eliminated a potential competitor on the Rødby-Puttgarden route and strengthened the joint dominant position of DSB and Deutsche Bundesbahn on that route. The effects of the refusal were further compounded by the refusal to authorise the construction of a new port. Both refusals were by the Danish Transport Minister and constituted state measures within the meaning of

Article 90 as they placed DSB, a public undertaking to which the state had granted exclusive rights, in a position in which it could not avoid infringing Article 86. In other words, had the refusals been made by DSB they would have constituted Article 86 abuses: therefore, when made by the Danish authorities they constituted state measures in breach of Article 90(1), read in conjunction with Article 86.

The Commission was unconvinced by the Danish arguments that their refusal was justified by technical constraints and insufficient market demand and found that the requirement that there should be an effect on trade between Member States was clearly satisfied. Furthermore, the effect was appreciable given the high volume of traffic on the Rødby-Puttgarden route.

Having found an infringement of Article 90(1), the Commission considered whether the exception laid down in Article 90(2) would apply. Article 90(2) provides that the EC competition rules apply to public undertakings such as DSB that have been entrusted with public service obligations, but *only* insofar as the rules do not obstruct the performance of such tasks. On the facts of the case, the Commission concluded that DSB's tasks were to organise rail services and to manage the port facilities at Rødby. These tasks were not impeded by the application of the competition rules and, therefore, the exception could not apply.

The Commission therefore decided that the Danish Government had infringed Article 90(1) and required it, within two months of the decision, to bring an end to its prohibition on the Stena subsidiaries from building a new port in the vicinity of Rødby operating from the existing facilities.

In the *Elsinore* case, the Commission examined the Danish Ministry of Transport's decision to refuse access to Elsinore port[51]. The case started in 1992 when the Danish shipping line, Mercandia, complained to the Commission that the Danish Ministry for Transport had refused it permission to operate a car and passenger ferry service from the port. The Commission considered the refusal of access to the port facilities in Elsinore to be a breach of Article 86 and Article 90 as it limited competition on the Elsinore-Helsingborg ferry route and reinforced the dominant position held by ScandLines. The Commission clearly took the view that the Danish Shipping line, Mercandia should have been given access to Elsinore port even though that port had no spare capacity so that existing operators were forced to cede part of their terminal capacity. The ramifications of this forced divestiture in such circumstances can hardly be understated. It is a manifest demonstration of the extent to which the EC model of essential facilities has departed from its American counterpart, where feasibility is one of the fundamental criteria which must be met before the doctrine is triggered[52].

Following negotiations with the Commission, the Danish Government agreed to allow a competing ferry operator access to the port of Elsinore by 1 June 1996. The new ferry operator was to be chosen by public tender procedure. As a result of the settlement, the Commission considered that it was not necessary to adopt the threatened Article 90 decision[53].

The most recent case dealing with the application of EC competition law and the operation of sea port facilities concerned the port of Roscoff in Brittany. The Commission recently awarded interim measures against the Chambre de Commerce et d'Industrie de Morlaix ('Morlaix Chamber of Commerce') in Brittany on the basis of a prima facie breach of Article 86 with regard to allowing access to the port[54]. The port of Roscoff is, for the time being, the only port capable of providing adequate port facilities in France for ferry services between Brittany and Ireland, a market which accounted for around 100,000 passengers in 1994. At present the only operator on the routes between Ireland and Brittany is Brittany Ferries, a company partly controlled by the Morlaix Chamber of Commerce. The Irish ferry company, Irish Continental Group ('IGC') wished to gain access to this market commencing with the 1995 season when it was hoping to offer services between the Irish ports of Cork and Rosslare and Roscoff.

In November 1994, IGC applied to the Morlaix Chamber of Commerce for access to the Breton port with a view to commencing a service between Ireland and Brittany for the season beginning on 27th May 1995. Negotiations carried on until the following month with the parties having reached agreement in principle on IGC's access to the port having, it appeared, successfully agreed on sailing schedules and a number of technical issues. IGC then announced its new services in December 1994 and began to accept bookings for the 1995 season for the new routes between Cork and Roscoff and Rosslare and Roscoff. However, in January 1995, the Morlaix Chamber of Commerce stated that it wished to suspend existing negotiations and started to try and re-negotiate the deal that had been agreed in principle, in particular with regard to the proposed date for commencing operations. In response to this IGC filed a complaint with the European Commission concerning the difficulties it was encountering in trying to offer services on these routes and claiming, inter alia, to have already taken over 40,000 bookings for its Cork-Roscoff and Rosslare-Roscoff routes for the 1995 season. Negotiations resumed at this time and it was hoped that the Commission would not in fact have to intervene but no agreement was reached, again notably as to the date to commence operations.

On May 16th 1995 the Commission decided on interim measures against the Morlaix Chamber of Commerce. The Commission made a prima facie finding that the Chamber of Commerce had abused its dominant position as the operator of the port of Roscoff by refusing IGC access to the port facilities there, in violation of Article 86 of the EC Treaty. The Commission further stated that this behaviour amounted to a refusal to supply services since Roscoff is the only port in France with the installations needed for operating ferry services between Brittany and Ireland. On the basis of the Commission decision, the Chamber of Commerce was obliged to take the necessary steps to allow IGC access to the port of Roscoff by June 10th 1995.

The Commission's decisions in the above cases are of great significance for operators of major ports throughout the European Union. The fact that the operators of ports as small as Holyhead and Roscoff are subject to the essential facilities doctrine shows that factors such as the significance of the routes served by the ports, in relation to intra Member State trade or travel, may be decisive as to the requirement that dominance is established 'in a substantial part of the common market'. The question of whether the port is an essential facility will depend in all cases on the availability of alternative ports for the required traffic.

Conclusion

With the accumulation of Commission jurisprudence developing the doctrine of essential facilities, dominant undertakings in the provision of such facilities are clearly under a duty to act fairly and on a non-discriminatory basis. Whilst both the operator and operator/user of the facility must prove that their behaviour is free of any taint of discrimination, the burden is considerably higher on the operator/user. In this regard, any arguments as to objective justification will be subject to intense scrutiny. For example, in *Frankfurt Flughafen*, the findings of experts' technical reports revealed assertions relating to the possible harm of admitting competitors to the ramp to be unfounded. There was found to be sufficient parking space on the apron of the airport to accommodate competitors' equipment without significantly impairing the safety of operations.

In a wider sense, the doctrine of essential facilities has proved an effective tool in opening up a particular facility to competition by forcing the operator to permit access to new entrants (*Frankfurt/Holyhead II*) or to provide the facility to existing competitors on fair and equitable terms (*London European - Sabena, Aéroports de Paris, Holyhead I*). Where cases merit the Commission's intervention, the competition rules are evidently to the complainant's benefit and, in some cases, a speedy resolution can be achieved[55].

Finally, and as mentioned in the introduction, the case law and developments described in this paper are not only relevant regarding the application of the EC competition and regulatory rules to essential facilities in the transport sector, but will also have a major impact upon the way in which the UK competition and regulatory authorities apply the Chapter II prohibition contained in the Competition Bill to behaviour relating to essential facilities which would not fall within the scope of Article 86 but could, nevertheless, fall within Chapter II.

Acknowledgement

The assistance of Martin Bailey, Norton Rose, in the preparation of this paper is gratefully acknowledged.

Notes

1   See further the Commission's Notice on the definition of relevant market for the purposes of Community competition law (97/C 372/03, OJ [1977] C 372.
2   Cases T-68, 77-78/79, Re Italian Flat Glass: Societa Italiana Vetro v Commission [1992] 5 CMLR 302.
3   OJ [1992] L134/1 French West African Shipowners' Committees and OJ [1993] L 34/20 Cewal.
4   The Commission decision was upheld by the Court of First Instance in Joined Cases T-24/93,25/93,26/93 and 28/93 *Compagnie Maritime Belge Transport v Commission* [1997] 4 CMLR 273.
5   See Press Release IP/98/811 of 16 September 1998.
6   OJ 1973 L140/17.
7   Case C-18/93 Corsica Ferries Italia Srl v. Corpo dei Piloti del Porte di Genova [1994] ECR I-1783.
8   Case C-62/86 *Akzo Chemie BV v Commission* [1991] ECR I-3359.

9   Ryanair's complaint of predatory pricing against Aer Lingus on the Dublin-Birmingham route never reached the stage of a decision, as Aer Lingus withdrew the fares in question after a dawn raid. In *EasyJet v KLM* the Commission conducted a dawn raid and issued a statement of objections but EasyJet later withdrew the complaint, it is understood because the Commission had changed its mind on the dominance issue given the degree of competition that exists on the relevant market. Nevertheless, it is understood that while there was clear evidence of predation, the problem was market dominance in the absence of which there can be no Article 86 infringement. See further the *Aer Lingus* decision, OJ [1994] L 54. Furthermore, in *Compagnie Maritime Belge Transports SA v Commission*, the practices of using fighting ships sailing close to the dates of the competitor and jointly fixing rates below conference rates were deemed to be abusive behaviour in breach of Article 86, but not predatory pricing. In its Decision, *Cewal* OJ [1993] L34/2, the Commission specifically made clear that predatory pricing can only apply to practices carried out by single undertakings since it distinguished between '*a concerted decision by several undertakings forming, in this case, a shipping conference aimed at fixing, within the framework of a plan, a special price to remove a competitor, and the case already examined by the Commission and the Court of Justice of abusively low prices established by a single undertaking acting unilaterally, where it was necessary to distinguish between predatory prices and aggressive competition*'.

10   Case C27/76 *United Brands Company and United Brands Continentaal BV v Commission* [1978] ECR 207.

11   However, in Case C-179/90 *Merci conventionali porto di Genova SpA v Siderugica Gabriella SpA* [1991] ECR I-5889 the Court held provisions of Italian law which reserved certain port activities to designated undertakings to be in breach of Article 90, by inducing those undertakings to abuse their dominant position *inter alia* by demanding payment for services not requested or charging disproportionate prices.

12   Cases 6 and 7/73 *Instituto Chemioterapico Italiano SpA and Commercial Solvents v Commission* [1974] ECR 223.

13   OJ [1992] L96/34.

14   *Alpha Flight Services/Aéroports de Paris* OJ [1998] L230/10.

15   See the Commission's decision in *Volkswagen,* OJ [1998] L124/60 for the first example of the application of its new guidelines on setting fines. See also the TACA case mentioned above, where a fine of ECU 273 million was imposed.

16   Case C-7/97 *Oscar Bronner GmbH & Co. KG v Media Zeitungs- und Zeitschriften GmbH & Co. KG and Others*, Opinion of Advocate General

Jacobs delivered on 28 May 1998. The alleged abuse concerned refusal of access, or imposition of unreasonable terms for access, by a newspaper to its national distribution system.

17 *B&I Line plc v Sealink Harbours Ltd and Sealink Stena Ltd* [1992] 5 CMLR 255.

18 224 U.S. 383 [1912].

19 Page 409.

20 410 U.S. 366 [1973].

21 570 F. 2d 982, 992-93 (D.C. Cir. 1977).

22 Page 992.

23 597 F. 2d 1081 (7th Cir. 1983).

24 807 F. 2d 520 (7th Cir. 1986).

25 Page 540.

26 1987-2 Trade Cases 67, 741 (S.D. Fla. 1987).

27 See for example Daniel E.Troy's article in the *Columbia Law Review*, vol. 83 1983. pp. 441-487 at p. 441, *'Unfortunately, courts have not developed a consistent analytic framework for the essential facility context. No clear rules govern when the essential facility doctrine should be invoked. Nor is there a consensus as to what the doctrine requires once invoked.'*

28 OJ [1995] L216/8.

29 *FAG - Flughafen Frankfurt/Main AG*, OJ [1998] L 72/30.

30 OJ [1988] L317/47.

31 XXIIIrd Report on Competition Policy, point 223.

32 In this case, the Court took particular account of the volume of traffic in the port and its importance in relation to the overall volume of imports and exports by sea to and from Italy.

33 It is interesting to note here the position under Article 8 of the Ground-handling directive (see footnote 38 below), where Member States, who reserve the management of 'centralised infrastructures' - which are defined and clearly are essential facilities - to the airport's managing body as a derogation from the liberalising provisions, must ensure that such management is transparent, objective and non-discriminatory.

34 XXth Report on Competition Policy, point 38.

35 The Commission ended its infringement procedure begun in 1994 in relation to Athens airport having obtained certain improvements in the supply of ground-handling services, including improvements to the eastern terminal housing foreign airlines and appointment of a new operator for ground-handling services. See Commission Press Release IP/97/876.

36   Case C-311/84 *Centre Belge d'études de marché - Télémarketing (CBEM) v SA Compagnie Luxembourgeoise de télédiffusion (CLT) and Information publicité Benelux,* ECR 3261 [1985].

37   Ramp-side ground-handling activities consist mainly of the following activities: provision and operation of equipment for embarkation and disembarkation of passengers; transport of passengers to and from the terminal; crew transport; loading and unloading of baggage and cargo; transport, sorting and transfer of baggage; transport of cargo; cabin cleaning; toilet and water services; push-back/towing of the transport of cargo; cabin cleaning; toilet and water services; push-back/towing of the aircraft; provision and operation of equipment to carry out the above mentioned activities; fuelling and supply of catering.

38   Council Directive 96/67/EC on access to the ground-handling market at Community airports, OJ [1996] L 272/36.

39   Commission Decision of 14 January 1998 on the application of Article 9 of Council Directive 96/67/EC to Frankfurt Airport. OJ [1988] L173/32.

40   Commission Press release of 8 September 1998, IP/98/794.

41   The Commission in *Sea Containers v Stena Sealink* stated that *"An undertaking which occupies a dominant position in the provision of an essential facility...which refuses other companies access to that facility without objective justification or grants access to competitors only on less favourable terms than those which it grants its own services, infringes Article 86 if the other conditions of that Article are met.* [Emphasis added]."

42   *B&I Line v. Sealink Harbours Ltd and Stena Sealink Ltd,* [1992] 5 CMLR 255 and *Sea Containers v. Stena Sealink,* OJ 1994 L15/8.

43   Recital 41. The footnote to that recital lists a series of cases relied upon by the Commission in support of its analysis, in particular leveraging cases such as Case 311/84, *Telemarketing* [1985] ECR 3261, Case C-260/89, *Elleniki Radiophonia Tiléorasi* [1991] I ECR-2925 and Cases T-69/89, T-70/89, and T-75/89, *RTE v. Commission* [1991] II ECR 485.

44   Recital 42.

45   See recitals 29 and 31.

46   Recital 70.

47   Recitals 74 and 75.

48   740 F. 2d. 980.

49   OJ 1994 L55/52.

50   Recital 12.

51   XXVth Report on Competition Policy, p.120.

52   See above the four criteria set out in *MCI Communications.*

53 See further Commission Press releases IP/96/456 and IP/96/205.

54 Commission Press Release IP/95/492, May 16th 1995.

55 For example, in the *Roscoff* case, it is reported that only two months elapsed between the refusal of access to the port and the Commission's decision forcing the port operator to admit Irish Ferries. Similarly, in follow-up to its decision in *Frankfurt Flughafen*, the Commission responded quickly to put an end to the imposition of long-term contracts on users of the airport.

13 Essential Facilities and Transport Infrastructure

MELANIE FARQUHARSON

Introduction

Trevor Soames' paper has provided an exposé of the genesis and application of the essential facilities doctrine in EC law as it applies in relation to sea ports. There are two particular areas that I would like to cover in addition to the areas covered in Trevor's presentation. Those are the application of the concept of essential facilities in the rail sector and (briefly) the application of similar principles in relation to bus stations in the UK. As Trevor has pointed out in his paper, whatever principles apply in EC law will be incorporated into UK law once the Competition Bill is enacted (Clause 60 incorporates EC jurisprudence, although this in itself is a minefield). But of course the old law on market power is being retained in parallel in the UK so the existing UK case law will continue to be relevant.

First, though, I want to flag two points - one is very much a lawyer's point (but nonetheless valid for that). We know that the European Commission (particularly perhaps John Temple Lang) has been keen to apply the essential facilities concept in European law. However, we need to be clear that that application is an application of the existing legal instruments. The essential facilities concept is not a new area of law - it is the application of Articles 85 and 86, mainly 86, to a particular type of situation. Article 86 gives the Commission the tool to apply an essential facility doctrine to the extent that the essentialness of the facility equates to the dominance of its owner. But the case law indicates that dominance only exists in legal terms in relation to a market so you need to be pretty clear that there is a real 'market' for access to the essential facility.

The second point I want to make is that there is often confusion about whether the essential facilities doctrine applies where the owner of the facility is not active in the downstream market. Can a facility only be essential if the owner operates in the downstream market? Clearly not, the rail network may well be an essential facility - even in Great Britain. But in fact, if the essential facility concept in EC law is merely an application of EC law to a particular type of situation, the question of whether the owner of the facility is active in downstream markets is merely one of the factors in assessing whether the 'dominant' undertaking is behaving in a discriminatory way as between customers. Where the facility owner is present in the downstream market of course he has a clear incentive to discriminate. There is also normally the added complication that establishing the terms on which the owner deals with its own downstream operations is itself a difficult exercise, involving as it does an assessment of internal transfer prices which are capable of being 'manipulated' (or perhaps I should say in more neutral terms they will not necessarily reflect an open market distribution of the available profit as between upstream and downstream activities). This itself is assuming that there *are* any such 'transfer prices'. Where the upstream and downstream activities are within a single entity, there is not even a starting point for assessing whether downstream access is being offered to a third party on non-discriminatory terms.

Sometimes of course the discrimination will not be financial. The mental picture of B&I Ferries bouncing up and down in the wake of a departing Sealink ferry and B&I's embarking and disembarking passengers tumbling into the sea in the process clearly took the Commission a long way in its analysis of discrimination.

Rail Infrastructure

Let me turn then to the rail sector. The first point to make is that the EC has *by legislation* sought to prevent the essential facilities problem from arising. In particular, Directive 91/440 required the separation (at least in accounting terms) of rail infrastructure management and rail transport operations. So we now have to talk about infrastructure managers 'IMs' and railway undertakings 'RUs'. The Commission would like to have gone further and required structural separation. Some Member States have gone down that route (the UK (at least for Great Britain) has done so with some vigour) - the Swedes have done so, the Germans will have done so from the beginning of next year.

Direction 91/440 also gave rights of access to rail infrastructure to certain entities namely railway undertakings (ie the operational bit, rather than the infrastructure bit) engaged in the international combined transport of goods and international groupings (strange concept) of railway undertakings providing cross-border passenger transport. However, those rights were limited to access in the country of the members of the grouping and access to the infrastructure of other EU countries only for transit (not cabotage). The Directive did say however that such access had to be provided on non-discriminatory terms.

Of course our GB system goes much further. The right to apply for access to railway facilities under s17 of the Railways Act 1993 is not so limited as Directive 91/440. Under s17 an applicant for access can apply to the Rail Regulator who will essentially be the arbiter of the terms of such access, so that there should be no concern about discrimination.

The potential for Great Britain rail facility owners (not just Railtrack, but the operators of the stations and light maintenance depots amongst others) to fall foul of the essential facilities doctrine is thus curtailed.

Whilst we now do not find this regulatory structure particularly shocking, we should bear in mind that the concept of infrastructure/operational separation was (and in some cases still is) anathema to some Member States and therefore the Commission's efforts to build on 91/440 since its adoption have been met with severe political resistance.

The July 1998 Rail 'Infrastructure Package'

On 22 July 1998 the European Commission tabled three proposals for Directives which, together with a White Paper on common principles of a transport infrastructure charging framework, are commonly referred to as the 'infrastructure package'.

In the context of the essential facilities concept, the aims of the Capacity Allocation proposal - which has yet to be debated formally by the EU Council of Ministers - are:

- To ensure that all railway undertakings (not just those covered by Directive 91/440/EEC (i.e. undertakings engaged in international combined transport/groupings of railway undertakings)) have equitable and non-discriminatory access to rail infrastructure in other EU countries;
- To do so carefully defining the respective rights and obligations of rail undertakings and infrastructure managers when it comes to the allocation of access rights to a scare resource, namely capacity/paths on a rail network;
- To remove the conflict of interest inherent in the capacity allocation decisions being taken by an entity which also uses that same facility, in

other words by operating trains on the rail network (and which also sets the access charges, fixes timetables and regulates safety issues);

- To require the efficient use of that scarce capacity by promoting bidding processes for particular 'train paths' and by setting access charges on marginal social cost basis (i.e. those variable costs which reflect the cost of an additional vehicle/transport unit using the infrastructure plus flexibility to allow for higher costs recovery, inclusion of external costs and scarcity problems.

'Essential Facilities' in Rail Transport

The European Court of Justice (ECJ) has held that an undertaking holding the monopoly in the market for the establishment and operation of a telecommunications network infringed Article 86 where it, without any objective necessity, reserved to itself the neighbouring but separate market for the importation, marketing, connection, commissioning and maintenance of equipment for connection to that network, thereby eliminating all competition from other undertakings[1].

To some extent it could be argued that the provisions of the Capacity Allocation proposal represent a retreat from the position in *GB-Inno-BM* relating to the separation of the regulatory and operational functions of a dominant undertaking owning/operating an essential facility.

The Commission's legislative steps to liberalise access to facilities in the rail sector do not of course exclude the application of Articles 85 and 86. They cannot. Not all Member States have achieved even accounting separation between rail infrastructure and operations and even where there is such separation there remains in many cases a dogged antipathy to new entry at the operational level, much to the frustration of the European Commission. The field is ripe for the application of the essential facilities doctrine.

So, the situations in which the dominant, incumbent rail infrastructure owner could fall foul of Article 86 EC Treaty could include the following in particular:

- The incumbent, vertically integrated rail undertaking cuts off services (e.g. access to network; maintenance services; traction services) to the authorised applicant which are essential for the provision of competing rail transport services in that there are no other real or potential substitutes available[2];

- The incumbent, vertically integrated rail undertaking refuses to supply these services to the authorised applicant thereby preventing the introduction of a new, competing rail (freight/passenger) transport service which is in specific, regular and constant demand from potential customers[3];
- The refusal of the incumbent operator of a facility such as the Channel Tunnel to supply to the potential competitors of the specialised services the special locomotives and their crews in order to operate in such an environment[4];
- The incumbent, vertically integrated rail undertaking grants access to the network, maintenance, traction services only on terms which are less favourable than those which it grants to its own services[5];
- The incumbent, vertically integrated rail undertaking grants access to its services on condition that the authorised applicant contracts to purchase other, ancillary services which have no objective, commercially justifiable connection with the services which the applicant requires in order to operate the competing rail transport service[6];
- The incumbent vertically integrated rail undertaking refuses to develop or even maintain the rail infrastructure being used by the authorised applicant or sets excessively onerous conditions for the development/maintenance of the infrastructure.

The European Court of First Instance helpfully produced a ruling on this area on 15 September 1998, concerning European Night Services. It is rather a sorry tale in that the Commission was roundly criticised by the Court for failing to establish its case properly. The Court overturned the Commission's decision, but does not always say what the right answer would be. European Night Services was the joint venture between various railway companies (set up before structural separation was required incidentally) to operate night train services through the Channel Tunnel. It is worth mentioning at this point that it has since become apparent that the operation of such services is unlikely to be viable in the near future - so they may never happen anyway. It should also be noted, since this is relevant for the application of the essential facility concept, that the rolling stock required to run through the Tunnel and on to the networks of the various countries into which these services were intended to go is very specialised and, at the moment, expensive. In view of these factors it was recognised that none of the participating companies would have taken the risk of launching a night service alone.

So in the ENS case the parties (BR/EPS/EUKL, NS, SNCF and DB) notified their joint venture arrangement for clearance or exemption under Article 85 (or in fact the land transport equivalent) to the Commission. The Commission granted an exemption but limited the duration of the exemption to seven years (whereas the period of the financing arrangement for the rolling stock was 20 years) and imposed the condition that new entrants must be able to purchase from the notifying parties the same rail services as those parties, as parents, had undertaken to sell to ENS on the same technical and financial terms. Those services were the provision of the locomotive, train crew and path on the networks and through the Channel Tunnel. The Commission believed that ENS was a 'transport operator' (which seems to be something downstream of a railway undertaking) and that the parents all remained active on a market downstream from ENS, namely the market in necessary rail services.

The Court found that there was no real market in these 'necessary' rail services, because in the passenger sector there did not exist 'transport operators' as envisaged by the Commission. ENS was simply a joint venture grouping of railway undertakings. It should be noted of course that in *Magill*[^1] Article 86 was applied in relation to a refusal to supply which prevented a new market from developing. The Commission found that ENS's parents held dominant positions in the supply of rail services in their own countries, especially as regards special locomotives for the Channel Tunnel.

The Court said that in the context of a joint venture neither the parents nor the joint venture itself should be regarded as being in possession of infrastructure, products or services which are necessary or essential for entry into the relevant market unless such infrastructure, products or services are not interchangeable and unless by reason of their special characteristics - in particular the prohibitive cost of and/or time reasonably required for reproducing them - there are no viable alternatives available to potential competitors for the joint venture which are therefore excluded from the market.

Where the Commission had made life difficult for itself was in saying that ENS's competitors were other operators in 1. the business and 2. the leisure travel markets on the routes ENS was to serve: Brussels to Glasgow for example. Clearly, if you want to compete in that market access to the rail services the Commission described is not essential - there are plenty of competitors who operate other modes of transport. The Commission had acknowledged that the relevant markets in which ENS operated was wider than the markets for business and leisure travel *by rail*.

This serves to emphasise my earlier point that market definition is key to the application of the essential facilities concept in European law. Not only had the Commission (according to the Court) got it wrong by saying that there were 'transport operators' wanting to purchase the services of train crews and locomotives so as to run passenger trains. It had also failed to show in its reasoning that even the provision of access to the infrastructure was an essential facility for ENS's competitors (because those entities were busy flying or sailing). Had it set out the reasoning correctly, the Commission might have established the essentialness of access to the rail infrastructure.

As I have mentioned the ENS case was not an Article 86 - dominance - case but an Article 85 - co-operative agreements - case. The Commission sought to apply the essential facilities concept in the context of the criteria under Article 85(3) for exemption from the prohibition on restrictive agreements. One of those criteria is that the agreement should not eliminate competition in a substantial part of the products or services in question - which can mean that an agreement which creates dominance would not be eligible for exemption unless it were subject to conditions ensuring access. In some Block Exemption Regulations under Article 85(3) (e.g. computer reservation systems) the Commission has required third party access as a condition of exemption.

However, in ENS the Commission saw its application of the doctrine as an application of the criterion under Article 85(3) that the restrictions in the agreement must be indispensable to the achievement of the pro-competitive benefits the agreement will bring. The Court said that the Commission had not properly established that the agreement did have restrictive effects so the relevance of the issue of indispensability was not clear. Even if it had been, however, the Commission would have had to establish that the relevant services were necessary or essential to competitors of ENS. The Court noted that access to train paths was in any event secured by Directive 91/440. The arguments for the essentialness of locomotives and crew were not made out because it was possible that new entrants could have bought or hired them. The Court dismissed the idea of a train crew as an essential facility in a single sentence. It was clearly sympathetic to the argument that to require the parties to the joint venture to share the benefits of their co-operation with third parties without those parties having to bear any of the commercial risks involved amounted to an expropriation.

In many ways the *Night Services* Decision must surely represent the high water mark of the Commission's application of the essential facilities concept to rail, particularly if as I believe is likely, the Court follows the opinion of Advocate General Jacobs in the *Bronner*[8] case and establishes some limits for the application of the concept.

Buses

I promised that I would look briefly at the bus cases in the UK. It is worth noting that the Transport Act 1985 imposed a specific obligation on municipal bus stations to provide access on non-discriminatory terms, but general principles apply in relation to privately owned stations. First *Southern Vectis*[9] - which concerned access to the Newport Bus Station on the Isle of Wight where the incumbent and former monopolist, Southern Vectis, prevented access to its bus station by a new entrant. This was a Competition Act investigation referred to the MMC in 1988 but Southern Vectis gave undertakings to the OFT to avoid the reference.

The OFT's assessment of whether Newport Bus Station was an essential facility started from the position that because of historical factors passengers were likely to assume that all available bus services started and ended at the bus station and even if the new entrant engaged in advertising or marketing activities it was unlikely that this would have much effect on consumers' preconceptions. The OFT also concluded that it would not be practicable or feasible for new operators to construct their own bus stations - partly on grounds of the size of such an investment for a new entrant and partly because of the function of a bus station - like other transport facilities - as a hub where passengers can choose between competing services and change between services operated by different customers. This assessment of essentialness has attracted some criticism.

The undertakings given by Southern Vectis were to permit rival operators access for a period of 10 years on terms comprising a pro rata share of the operating costs of the station (including a return on capital).

In the Report the OFT mused on the fundamental conflict between the interests of competition and the commercial interests of a bus operator owning the only bus station in the area. It suggested that separation of infrastructure from operation and non-discriminatory access charging on an arm's length basis would be one way of resolving this conflict.

The second case is *Mid Kent/West Kent*[10] which was referred to the MMC under the Fair Trading Act's scale monopoly provisions. The MMC's 1993 Report found that Maidstone & District Motor Services Limited ('MD') had imposed limitations on access to its Pentagon Bus Station in Chatham which were against the public interest. The undertakings given following that report were less strict but more detailed than those given by Southern Vectis. For example, MD was only required to allow an aggregate of 10% of the capacity of its bus stands for use by its rivals.

Notes

1   Case CO18/88 *Régie de Télégraphes et des Téléphones -v- GB-Inno-BM* [1991] ECR 3261.
2   Joined cases C-241/91P and C-242/91P *RTE and ITP -v- Commission of the European Communities* [1995] ECR I-743; most recently applied in Joined Cases T-374/94, T-375/94, T-384/94 and T-388/94 *European Night Servicews, Eurostar and others -v- Commission of the European Communities,* judgement of the Court of First Instance of 15 September 1998, not yet reported.
3   *Ibid.* The Austrian 1988 Rail Transport Law *(Eisenbahnbeforderungsgesetz)* specifically prohibits ÖBB from unjustifiably refusing to enter into contracts to allow access to its network.
4   *European Night Services,* paragraphs 212-216, footnote 9; although in that case the Commission's original Decision was annulled by the Court partly because the Court agreed that because European Night Services's market share of the relevant, intra modal market for business/leisure travel was low, there was no dominant position and consequently the refusal of access to these essential facilities could not constitute an abuse of such a dominant position.
5   Commission Decision 94/19/EC of 21 December 1993 relating to a proceeding pursuant to Article 86 of the EC Treaty, *Sea Containers -v- Stena Sealink* (OJ 1994 L15/8).
6   Case 311/84 *CBEM -v- CLT and IPB* [1985] ECR 3261.
7   *RTE and ITP -v- Commission.* Case C241/91 and C242/91, [1995] ECR I-0743.
8   Opinion of 28 May 1998, Case C-7/97 *Oscar Bronner GmbH & Co KG -v- Mediaprint Zeitungs-und Zeitschriftenverlag GmbH & Co KG and Others.*
9   OFT Report 17 February 1988.
10 MMC Report Cm2309-1993.